EXPERIMENTS IN ASTRONOMY FOR AMATEURS

EXPERIMENTS IN ASTRONOMY FOR AMATEURS

RICHARD KNOX

ST. MARTIN'S PRESS
NEW YORK

To Pat, David and Stuart,
for their help and patience.

St. Martin's Press, Inc., 175 Fifth Ave., New York, N.Y. 10010

Library of Congress Catalog Card Number: 75-34748
First published in the United States of America in 1976

Contents

	Page
Foreword	7
1 The moving Earth	9
2 Time by the Sun	24
3 Time by the stars	46
4 Stars and constellations	57
5 Models of the celestial sphere	78
6 Measuring the sky	94
7 Measurement on the celestial sphere	117
8 The wandering stars	136
9 The double planet	162
Postscript	190
Appendix A (Answers to exercises)	191
Appendix B (Mathematical methods and tables)	
Symbols and abbreviations	192
Solutions to astronomical and plane triangles	192
Sidereal time at 00h GMT throughout the year	194
Sun's declination, and the equation of time	195
Mean elements of planetary orbits, 1971 Jan. 0.5 d	196
Summary of data and formulae for planetary and lunar position calculations	197
Number of days through the year	199
Some star names, positions and magnitudes	200
Further Reading	201
Index	202

'All this is a dream. Still, examine it by a few experiments. Nothing is too wonderful to be true if it be consistent with the laws of nature, and in such things as these, experiment is the best test of such consistency'

Michael Faraday (1791–1867)

Foreword

The student of astronomy, of whatever age, can pursue his interest in a number of ways. But many newcomers to the subject feel a sense of frustration as they begin to learn a little about our solar system and the stars, because they have a very natural desire to find out some of the facts for themselves. Many would-be amateur astronomers lose interest at this point, without realising that astronomy was a flourishing science for thousands of years before the telescope was invented. Much of our basic astronomy evolved before Galileo and his contemporaries turned telescopes towards the stars, particularly those matters concerning the day to day and year to year behaviour of the stars and planets.

Even many experienced amateur astronomers today fail to understand properly the workings of the sky above them. So this book is intended to enlarge the realms in which both existing amateurs and newcomers to the subject can be interested. Most of all, it is intended to show how much there is to do, and this is one of the most effective and enjoyable ways to learn anything. Most of the book consists of experiments and exercises which use very simple devices that the amateur can make for a few small coins, and none which require an expensive astronomical telescope. But experiment and participation is what this book is all about and to get the best out of it the reader must carry out as much of the practical work as possible.

In general, the book has been written for the observer in the northern hemisphere since to draw attention to the reversal of east and west and north and south on every occasion would be tedious. However, where there are important differences in the techniques, attention has been drawn to them, so the southern observer will have no difficulty in carrying out the experiments. It would nevertheless be a good idea for the southern observer to get a good star atlas to cover the constellations south of the equator not mentioned in this text.

I would like to thank Dr J. G. Porter PhD FRAS for his invaluable assistance in reading the manuscript and making many constructive comments and suggestions for its improvement, and Mr J. L. White FRAS for his help and encouragement. Not least, I would like to thank Mr M. G. Clarke for his stalwart efforts in checking the contents and my typing, and for participating in many of the experiments described.

Richard Knox
Walton-on-Thames, 1974

I
The Moving Earth

The third planet of the solar system, counting from the Sun, is the Earth. From this planet Man has watched the sky for thousands of years. The ancient astronomers, particularly those of Egypt in the centuries before Christ, acquired a very detailed knowledge of the behaviour of the sky even though their view of the universe was very different from ours. One reason for the difference in views was that it was naturally assumed that the Earth was in the centre of the universe, because the Sun, Moon and stars could be seen going round the Earth.

Of course we now know that the Earth turns on its axis, so that the sky only appears to rotate above us, but there are still many peoples of 'The Third World' who, like the ancient astronomers, can see perfectly well that the sky turns about the Earth. If you had to prove that it was not so you would have an almost impossible task. Without the benefits of an advanced education and the accumulated knowledge of centuries which you take for granted, you could not show whether the system of planets in which we find ourselves is geocentric (Earth-centred) or heliocentric (Sun-centred). Indeed, for many practical purposes it does not matter if we use the geocentric system, since for the calculations and measurements we shall make it is quite suitable, and easier in many cases.

Daily Rotation of the Earth

First, we can investigate the daily rotation of the Earth (or sky) by watching the Sun. The Sun, like all the objects in the sky, makes one rotation about the Earth (as far as we are concerned) once in 24 hours. This ignores the annual motion of the Earth around the Sun which we will study later. The Sun rises in the east and sets in the west, the actual point of rising and setting varying through the year. For the moment, we will follow it across the sky once it has risen. A convenient way to do this is to watch the movement of a shadow across the ground. On a day which seems set fair, place a straight stick in the ground well away from any buildings or trees which could cast a shadow across the stick later in the day. Make the height of the stick a round number, say 150cm, and fix it as vertically as you can.

Starting as soon in the day as possible, place a small peg in the ground at the end of the shadow at regular intervals; once an hour on the hour is ideal. Be careful to note the time at which you place the first peg and to note any hour you may have missed, due, for example, to the Sun passing behind some clouds.

While the experiment is in progress hold a magnetic compass above the stick and note a distant object due south. You could also prepare a table in which to record the results, arranging them in columns with the following headings:

1	2	3	4	5	6	7	8
Date	Time (GMT)	Shadow length (cm)	Angle with previous shadow	distance peg-to-peg	Alt	Az	Sun's angle

You will then be able to complete columns 1, 2, 3 and 5 as the test proceeds. The others involve some calculations to complete. But you may prefer to measure the angle between each shadow and the previous one with a protractor rather than calculate them and put the result in column 4. If you repeat the experiment on several occasions throughout the year you will find some interesting differences between the sets of results.

The first observations to make, by standing on the shadow side of the stick and lining up the stick with the object to the south, is at what time the Sun's shadow points due north, that is, when is the Sun due south? The answer should be fairly close to 12 o'clock. British observers must be careful to subtract an hour from the clock time during British Summer Time to obtain Greenwich Mean Time, and get into the habit of using the 24 hour clock which we will employ in this book from now on.

From the table of results, firstly work out the angle between shadows to put in column 4 by scale drawing, or by using the cosine formula (this is given in appendix B). The *altitude* of the Sun is the angle to the horizontal of a line joining the Sun to a point on the ground. Fig 1 shows that this is the angle between the line from the end of the shadow to the top of the stick and the ground. It can be found by scale drawing, or by the simple solution of the right-angled triangle (see appendix B).

We see that the altitude of the Sun is changing all the time, except close to noon when it is fairly constant and at a maximum. From the successive shadows it appears as if the Sun moves *across* the sky faster close to noon than earlier or later in the day. The direction horizontally, for example

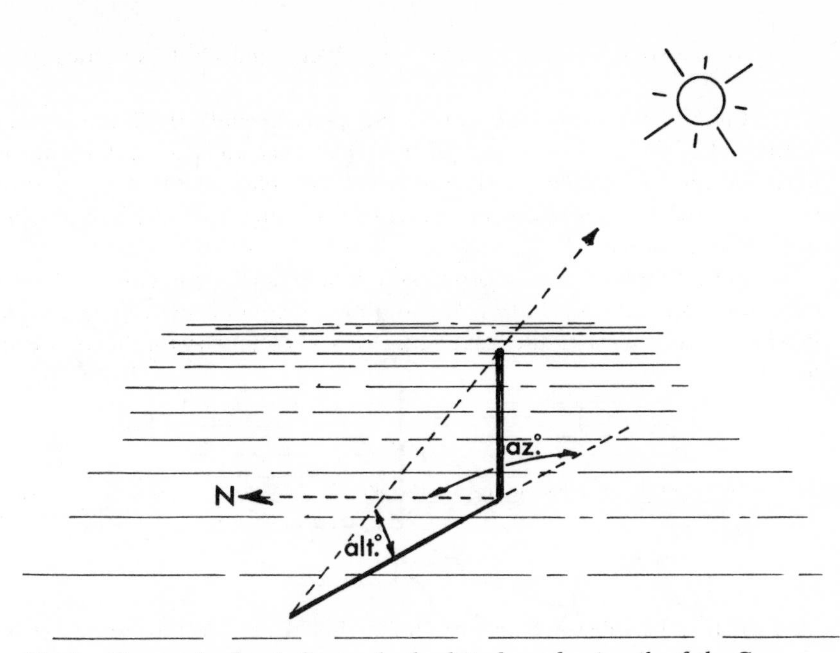

Fig 1 How a shadow indicates both altitude and azimuth of the Sun.

east, south-east, south, etc, ignoring the height (or altitude) is called the *azimuth*. Thus we have shown that the Sun is constantly changing both its altitude (height) and azimuth (direction).

The results also show that the Sun crosses the sky from east to west, reaches its highest point at about 12h, the middle of the day, and since we know that the motion is in fact due to the Earth's rotation, the Earth must turn on its axis from west to east. Because we turn towards the east as the Earth rotates, the Sun rises each day in Japan before India, in India before Europe, in Europe before the USA and so on.

Because both the altitude and azimuth of the Sun are changing the whole time, the actual angle through which the Sun passes each hour, which is what we want to put in column 8, must be calculated using both co-ordinates. Fig 1 shows how the angles of altitude (alt) and azimuth (az) can be measured from the shadow position. We can calculate the azimuth to go in column 7 of the table by making a scale drawing of each shadow. Alternatively, by measuring the sides of the triangle formed on the ground by the stick, the noon peg and the peg for each of the other hours in turn, we can find the Sun's direction east or west using the cosine formula.

The azimuth is usually expressed as the angle measured from the

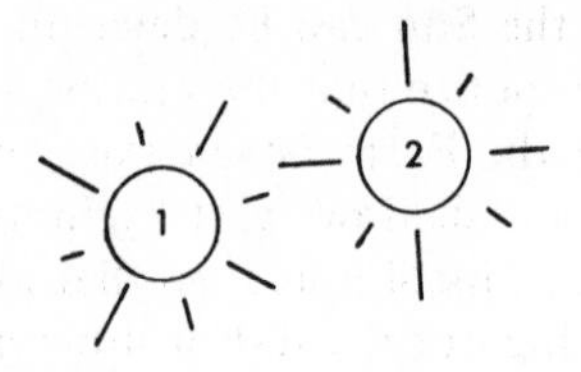

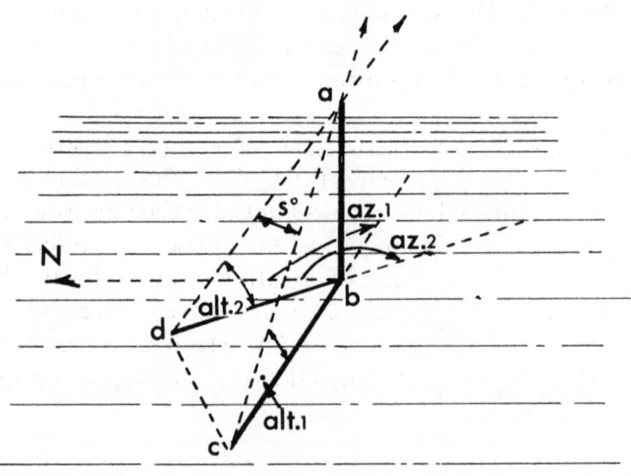

Fig 2 With two measurements of both altitude and azimuth at times T_1 and T_2, we find the actual angle S through which the Sun has passed during this interval.

northerly direction through east, south and west, in that order. Due east is azimuth 90°, for example, and south west is azimuth 225°. In some cases, however, azimuths may be expressed in some other stated way; for example south west could be described as 45° west of south.

Fig 2 shows how two successive positions of the shadow, with their associated altitudes and azimuths, can be used to find the angular motion of the Sun across the sky during that hour. By scale drawing, or trigonometry, we can work out the lengths *ac* and *ad* using the triangles *abc* and *dba* respectively. The distance between each peg, *cd*, has been measured during the experiment, so we can now draw the triangles *adc* for each hour, then measure or calculate angle S, which completes the table.

Now we must anticipate the results of this experiment before we can continue, and so if possible you should have completed the measurements before reading on. You will find that the Sun turns through the same angle each hour and that the angle will be close to 15°. In chapter seven we shall examine this question in more detail to see how the apparent angular motion of the Sun can be determined exactly, but for the moment it is sufficient to notice that the motion is uniform. It is due, of course, to the rotation of the Earth on its axis, and we would obtain similar results if we could plot shadows cast by individual stars. The Earth rotates once, through 360°, in 24 hours, so that in 1 hour it will turn through 15°.

By carrying out the test at different times of the year we will find that the Sun is seldom due south at noon exactly; it is seldom at the same altitude at the same time of day; and it rises and sets at very different points on the horizon. The Sun, in fact, although it is the basis of our time measuring system, is a rotten time keeper. It's the Earth which is really to blame, of course, and we shall return to this later to see what causes these peculiarities.

Inevitably, before Man learned to measure time by instruments, he regulated his life by the rising and setting of the Sun, so that it was only natural to measure time in days, each equal to the time the Sun took to go around the Earth. Since the time between sunrises or sunsets varied so much through the year in the temperate latitudes of the Earth, the length of the day was fixed at the average interval between successive noons, or midnights, and is called the mean solar day. Mean solar time is the basis of Greenwich Mean Time, and through the year sundials can be apparently fast or slow, by as much as 17 minutes.

The Earth's Axis

The Earth's diurnal (daily) rotation is about an axis which passes through the north and south poles (the poles of a planet are the points where the axis of rotation meets the surface). The huge mass of a planet spinning on this axis makes a very effective gyroscope, and just as the axis of a gyroscope is fixed in space, which is the basis of the gyroscopic compass, so the axis of a planet or of the Sun itself, is fixed in space. You can demonstrate this effect very convincingly next time you are taking the wheel off a bicycle. Set the wheel spinning while holding it by the axle, so that the wheel is spinning between your arms with the wheel vertical. Now turn the axle through 90° while the wheel is still spinning and you will feel a considerable resistance to the change.

As well as its diurnal rotation, the Earth is in orbit around the Sun, so that if the axis of diurnal rotation (the line joining the north and south poles) were at right angles to the plane of the Earth's orbit, the Sun would be exactly overhead at the equator all the year round, and would reach the same maximum altitude at noon each day when it was due south, from any given latitude. At the poles, the Sun would be perpetually on the horizon and would change its position only in azimuth.

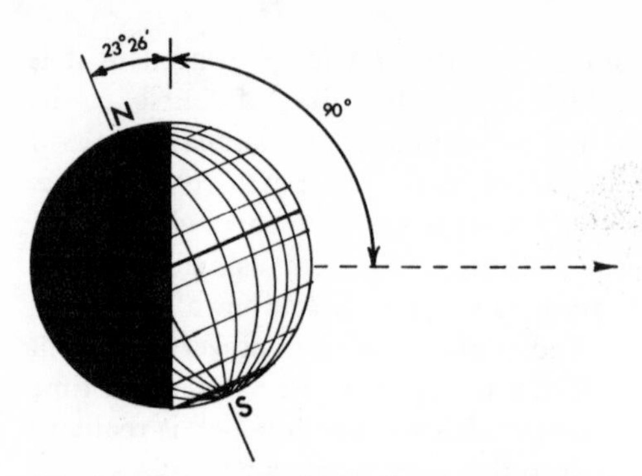

Fig 3 The axis of the Earth is at 23½° to the plane of the Earth's orbit. This shows the northern winter (and hence southern summer) solstice.

The Earth's axis is actually inclined to the plane of the Earth's orbit by just under 23½° as shown in Fig 3. In the position shown in this diagram it can be seen that, as the Earth turns on its axis each day, the north pole will be in perpetual darkness while the south pole will enjoy perpetual daylight. If you study this drawing closely you will see that the length of the equator in daylight is exactly the same as in darkness so that, despite the inclination of the Earth's axis, the equator has equal day and night. But half way from the equator to the north pole there is much more of that circle of latitude in darkness than in light.

Fig 3 shows the northern winter *solstice*, at which northern latitudes experience the shortest hours of daylight while the southern latitudes enjoy the summer solstice, when they receive the maximum length of daylight. The solstices occur at about December 22, and June 21. At the winter solstice in the northern hemisphere we can see that even 23½° from the north pole the Sun has not quite risen, even though it is due south at one point on this circle. Similarly, the Sun has not quite set at the opposite point on the globe, where it is midnight. These latitudes are the Arctic and Antarctic Circles respectively. In addition, we can see that the Sun is exactly overhead at a point 23½° south of the equator. This latitude is called the Tropic of Capricorn, which marks the farthest south that the Sun can be exactly overhead. There is a corresponding latitude of 23½° north called the Tropic of Cancer. The names are taken from the stellar constellations in which the Sun appeared centuries ago when the constellations were first named. In modern times the Sun is in adjacent constellations to these in mid summer and mid winter, because of a phenomenon called precession of the equinoxes. We will return to this later.

When the Earth is on the opposite side of its orbit to that shown in Fig 3 it will be the middle of the northern summer and the middle of the

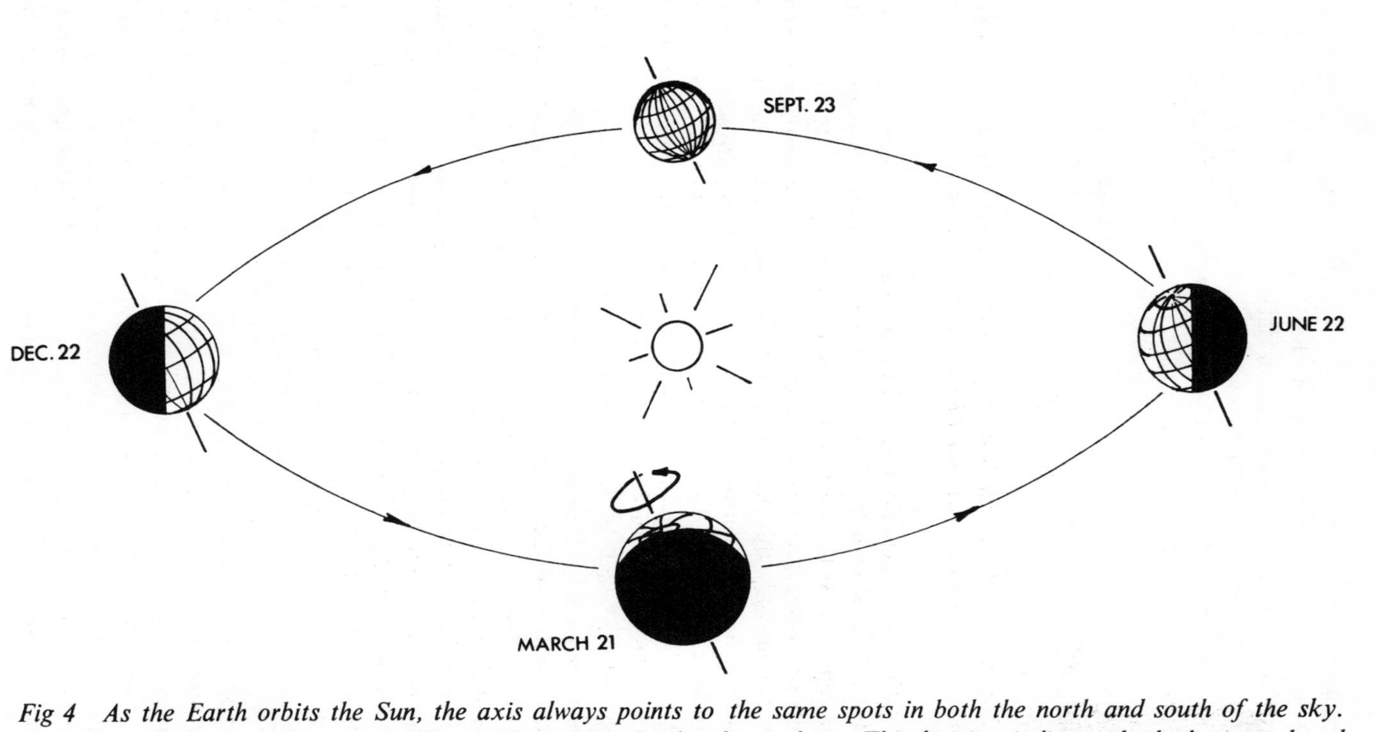

Fig 4 As the Earth orbits the Sun, the axis always points to the same spots in both the north and south of the sky. When a pole is inclined towards the Sun, it is summer in that hemisphere. The drawing indicates both the annual and diurnal rotation of the Earth.

southern winter. The picture would be very similar to Fig 3, except that the Sun would appear on the left of the Earth in the diagram and the illuminated half of the Earth would be in darkness and vice versa. The Sun would be overhead on the Tropic of Cancer, the north pole would be in perpetual daylight and the south pole in perpetual night. Half way between these two positions of the Earth in its orbit, if we still use the view of the Earth as shown in Fig 3, the Sun would be immediately behind the Earth in one case and in front of it in the other. In both positions, the sun light will be just reaching both poles. At these times, called the *equinoxes*, all parts of the Earth (except the poles themselves where the Sun is so low that in theory the horizon cuts it in half all day) experience equal daylight and darkness. The equinoxes occur at about March 21 and September 22. Another important point to note is that it makes no difference if the Sun moves around the Earth or the Earth goes round the Sun, because the axis keeps pointing to the same point in space wherever the Sun is in relation to the Earth, as shown in Fig 4 which illustrates all the solstices and equinoxes.

Parts of the Earth where the Sun's energy is reaching the surface at an oblique angle will be colder than where the energy reaches the surface more directly, hence the continuously changing seasons at latitudes away from the equator.

Parallax

The Earth is 150 million kilometres (93 million miles) from the Sun, so that at any date it is 300 million kilometres from its position 6 months earlier. In other words, we are looking at the stars from a very different position in space in mid winter compared with mid summer. Now try a very simple experiment. Close one eye and hold a finger in front of you as far away as you can. Now hold up another finger on the other hand, closer to your eye, and line it up with the first, still with one eye shut. Without moving your fingers at all, change the eye that you are looking at them with. The fingers are no longer in line because you have changed your viewpoint by a distance equal to the separation of your eyes, about 6cm. This effect is called *parallax*, and it could affect our view of the stars from our 300 million kilometre baseline.

Even the nearest star to our Sun, a star in the southern skies called Proxima Centauri is too far away to show any parallax that we would notice. Proxima is 4·2 light years distant, which means it is at the distance that light travels in 4·2 years. Since light travels at about 300,000 kilometres per second (186,000 miles a second), this puts the star about 40 million million kilometres (over 24½ million million miles) away. This star moves in relation to the more distant stars in the sky behind it by less than one four thousandth of a degree as seen from opposite positions in the Earth's orbit! To measure very small angles like this we divide the degree into 60 minutes of arc, written 60′ or 60 arc min, and each arc min into 60 seconds of arc, written 60″ or 60 arc sec. The variation in Proxima's position is about 1 arc sec.

Returning to Earth, we see that at all times of the year the stars appear to remain fixed relative to one another, and we cannot detect any change without very sensitive instruments. We can regard the stars as a fixed background to all we see in the sky and we can consider them, as ancient astronomers did, as being on a celestial sphere in the centre of which the Earth turns. On this sphere will be a spot to which the north pole points, and another where the south pole points—both poles pointing to these spots on the celestial sphere all day and every day. But remember that, because of the enormous distances to the stars, their positions appear the same wherever we are on the Earth, except for the purely local effects of how high each star is at a given time, and in what direction it is. Accordingly, if we could draw a line to a star from the Earth's equator, and other lines from each pole to the star, all three lines would be parallel.

Latitude and Longitude

Latitude is a measure of how far a point on the Earth is from the equator. We can drawn an imaginary line around the Earth on which all points would make the same angle to the plane of the equator at the centre of the Earth, as shown in Fig 5. This shows how three latitudes, 60° and 30° north (of the equator) and 45° south are fixed. Each circle of latitude is of a different diameter, getting larger the closer we are to the equator.

Longitude is measured from a line joining the poles and passing through a famous spot on the ground in the middle of the old Greenwich Observatory. This arbitrary point was fixed in the early days of the observatory, founded by Charles II to provide mariners with an accurate almanac with which they could determine their longitude, that is the distance east or west from England. Since the stars' positions were measured from Greenwich, this was a natural choice for the 'base line' of 0° longitude.

The lines of longitude, also called meridians, pass through the poles and so cross each other at both poles. The angle between any two lines of longitude at either pole is the difference in their longitude, as can be seen in Fig 6. This shows a view of the Earth from immediately above the north pole, with 0° longitude passing through Greenwich. On the opposite side of the Earth this longitude becomes 180°, since this is the angle it makes with 0° longitude at Greenwich, east or west. The drawing also shows

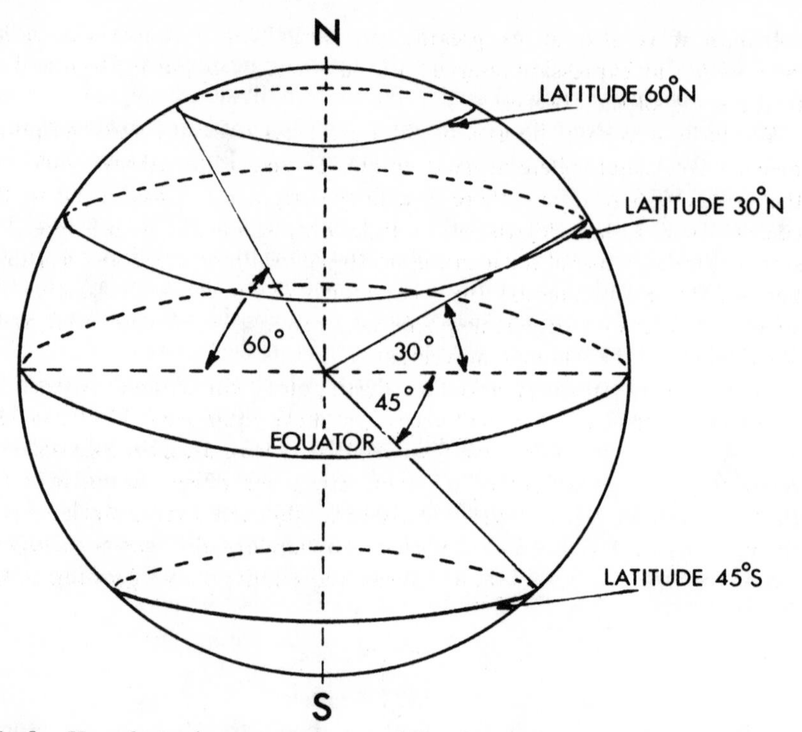

Fig 5 How latitudes are determined by the angles at the centre of the Earth.

another meridian for longitudes 60° west and 120° east, each of which make these angles with the 0° line.

A meridian on the surface of the Earth is a circle which passes through both poles. However, the term 'meridian' really comes from two Latin words meaning 'mid-day', and this is closer to its most important use in astronomy.

The *meridian,* or strictly, the observer's meridian, is a circle across the celestial sphere which passes over the observer's head, through the north and south points on his horizon, and through the celestial pole. When we say that the Sun is 'on the meridian' it means for observers in the northern hemisphere that it is due south and hence it is approximately the middle of the day as measured by our local mean time. But any other body can be on the meridian, and this normally means that it is due south, although some stars can cross the meridian under the pole, when they are due north. For southern hemisphere observers, objects on the meridian are due north, unless they are under the pole. When an astronomical body is on the meridian it is also at its greatest altitude. This is sometimes called *culmination,* an expression reserved for the object's crossing of the meridian to the south of the observer.

We have seen that the axis of the Earth, extended through the poles, meets the celestial sphere at fixed points. These are the celestial poles. In the northern hemisphere there is a fairly bright star quite close to the celestial pole. Called Polaris, or the Pole Star, because of its position, this star is extremely useful for locating north wherever we are in the northern hemisphere. Unfortunately, there is no such star to mark the south pole, so that in the southern hemisphere we must use other stars and make allowances for their hourly movement.

Polaris has not always marked the north pole so conveniently, due to the effect mentioned earlier of the 'precession of the equinoxes'. If we consider an ordinary toy gyroscope again, we see that as it rotates, the axis makes a relatively slow movement of its own. The same effect occurs with the Earth's rotation. The axis very slowly turns through a large circle so that the positions of the poles on the celestial sphere are also slowly changing.

The positions of stars were first measured relative to the position of the

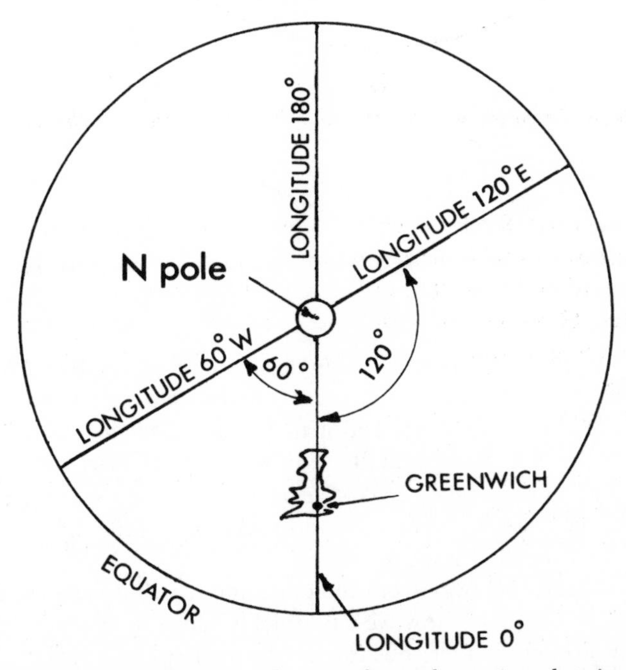

Fig 6 How the angle between the meridians determines longitude east of west of the Greenwich meridian.

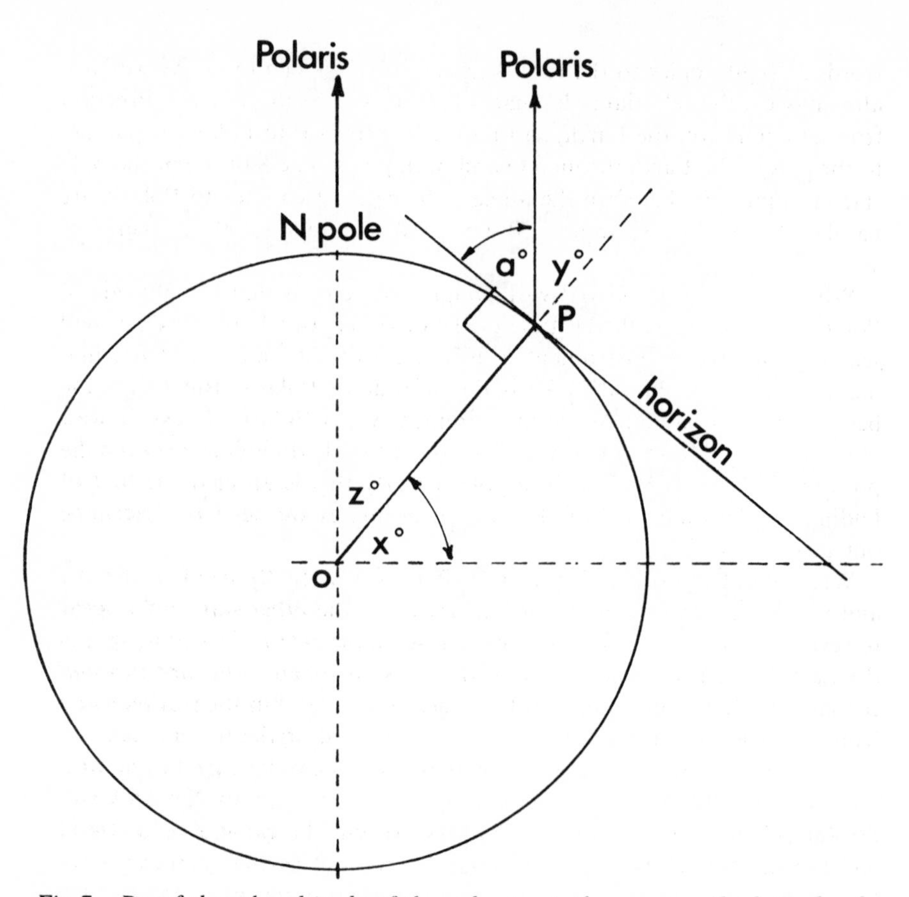

Fig 7 Proof that the altitude of the pole star is the same as the latitude of the observer.

Sun at the spring equinox when, many centuries ago, the Sun would be in the constellation of Aries. This point on the celestial sphere was then named the *first point of Aries,* and it is still so called even though, due to precession, the Sun is now in the constellation of Pisces at the spring equinox. But since it takes 26,000 years to make one complete cycle of precession, the change is very slow and makes very little difference to any measurements we shall want to make.

Polaris is nearly 1° away from the pole, and is still getting closer due to precession. We shall see how we can quickly judge its position relative to the pole in a later chapter, but for the moment we will ignore the small angle. Fig 7 shows an observer at latitude $x°$ at point P on the Earth. The horizon at P is the line tangential to the surface of the Earth, in other words at right angles to the radius joining P to the centre O. As we have already seen, the celestial pole, that is Polaris, will be in the same direction from any point on the Earth, so that the line from P to Polaris is parallel to the axis of the Earth through O and both poles. The altitude of the pole star at point P is therefore the angle a. Since the two lines to Polaris are parallel, the angle z = angle y. But $z = 90-x$ and $y = 90-a$, therefore $x = a$.

What we have just proved with simple geometry is that the altitude of Polaris is the same as the latitude of the observer. In Chapter six we shall see how to make an instrument to measure altitude, and one of the first measurements we should make is the altitude of Polaris. But to get the best result we must make the measurement when Polaris is east or west of the pole, not above or below it, because of that 1° difference between the position of Polaris and the true pole. Clearly this is an easy method of finding our latitude, one of the two co-ordinates we need to determine our position on the Earth.

What about the other stars? Since Polaris is virtually fixed in the sky and yet the celestial sphere is turning above us, the other stars will appear to revolve around the pole in circles. Those on the *celestial equator*, that is the Earth's equator projected on to the celestial sphere, will turn through the same angle in a given time as those near the poles, but the further away from the pole they are, the longer the arc they will make to our eyes.

A spectacular experiment to show these points can be carried out with a camera. Load your camera with a fairly fast film—about 200 ASA will do—and adjust the aperture to about f4. Mount the camera on a tripod and use a cable release with a locking device on it so that you can leave the camera with the shutter open. On a moonless night, place the camera in a spot shaded from any artificial lights and point it at the pole. Open the shutter and leave it for an hour. Try a number of exposures for both shorter and longer periods, and at different apertures. The camera must be perfectly still, and it should be sheltered from any wind. The focus must be at infinity. Remember to make a careful record of your exposures. You should repeat the experiment on different parts of the sky. Try close to the eastern or western horizons, if there are any bright stars close to them, and try the brighter constellations. The resulting photograph shows the path of the stars during the exposure, called a star trail.

Star trail photography is an extremely useful exercise: it helps to give a very clear awareness of the motion of the sky above you; it helps to teach you the individual stars and constellations fairly painlessly; it brings experience in night photography both for astronomical and other purposes; and photographs of the pole will show the little arc made by Polaris because of its small distance from the pole.

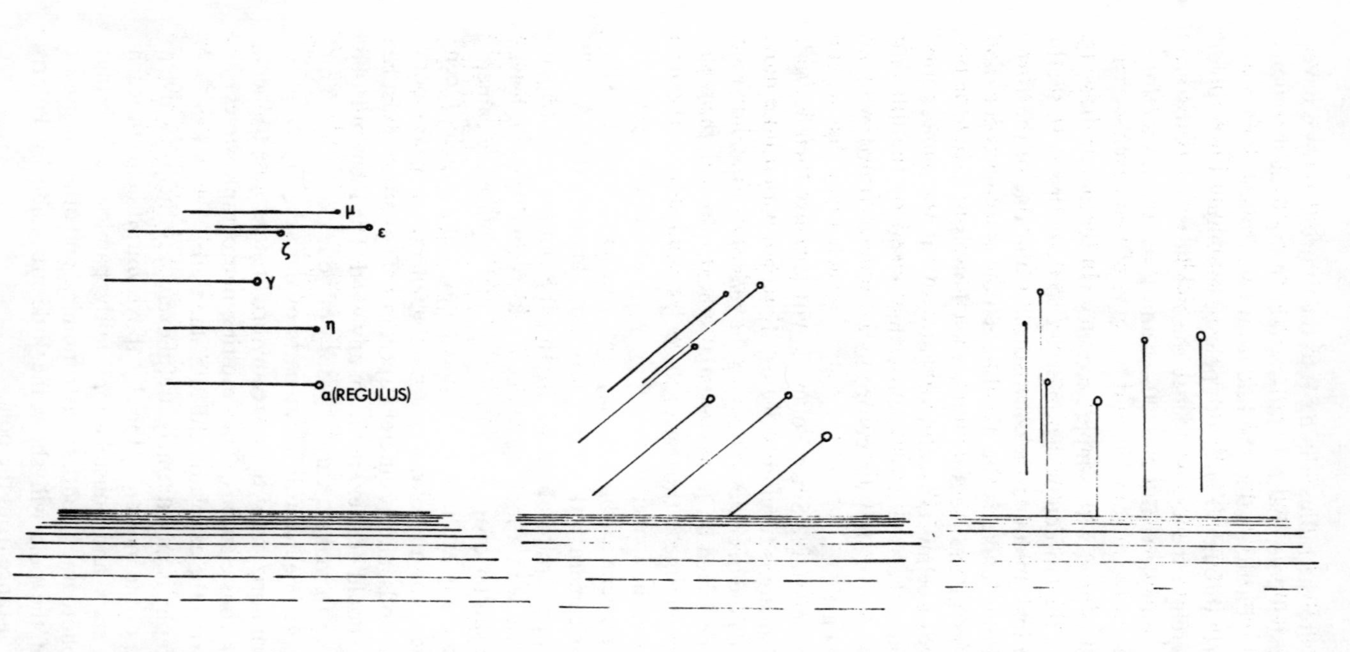

Fig 8 An impression of star trails made by some of the stars in Leo as seen (left) *at the North Pole,* (centre) *from London, and* (right) *at the equator.*

If you could visit the north pole and the equator and take star trail photographs of the same constellation you would see at once the different direction of the motion of the celestial sphere due to your latitude. Since few of us get such an opportunity, the result is illustrated in Fig 8 for the backward question mark sometimes called the Sickle, which is the head of the Lion in the constellation Leo.

On the left we are at the north pole, and the stars of Leo never set here; neither does the Sun during the northern summer. In the centre drawing, we are at latitude $51\frac{1}{2}°$ north and we can see that at the start of the 'exposure' Regulus had not yet risen. On the right of the drawing we have an impression of Leo rising at the equator. In this case, Regulus had already cleared the horizon at the beginning of the 'exposure', but the two lower stars had not. Each of the three are drawn for the same period of time, clearly showing the different altitudes attained by the different stars relative to each other at the different latitudes. Armed with an almanac which gives the positions of these stars in such a way that we can work out their distance from the pole, we could measure their altitude at a given moment and find our latitude from that. So, with a little more work, we can find our latitude without seeing the pole star. First, however, we need to look at how time is measured for astronomical purposes, and as the basis of time depends on the Sun we shall devote the next chapter to this star.

EXERCISES ON CHAPTER 1

1 In which direction does a shadow point at 12h? At what time is the shadow shortest? In which direction does a shadow point at the last moment before the Sun touches the horizon on March 21? A shadow of a vertical 2m pole is 2m long; what is the altitude of the Sun? At 18h the Sun is seen from London to be on the horizon; at what time did it rise, and what is the date?

2 The angle between a bright star and Polaris is measured at 18h on July 28 to be 90°; what will the angle be at 23h 30m on August 16? When the same bright star is due south it is found to have an altitude of 30°; what is the latitude of the observer? At the moment the star is due south, at what latitude will it be overhead?

3 A star in the northern sky is observed at an altitude of 30°, and 2 hours later it is due south at an altitude of 30°; what is the latitude of the observer? At a different location, a star is observed due east at 21h reasonably high in the sky, and later it is seen to be overhead; what is the latitude of the observer? If the star was overhead at midnight 00h 0m, at what altitude was it at 21h?

2
Time by the Sun

If you study the sunrise and sunset times in your diary, you will find that after the winter solstice, about December 22, the Sun rises at about the same time for several days; and indeed instead of starting to get earlier, sunrise gets a little later after the solstice. It is not until about January 6 that the sunrise starts to get earlier again, nearly two weeks after the solstice. Meanwhile, sunset has started to get later some days before the solstice, having reached its earliest in the afternoon on about December 13, almost two weeks before. The net result is that the Sun is above the horizon for the shortest total time about half way between December 13 and January 6.

With this strange behaviour, how can we expect the Sun to be half way between rising and setting exactly 24 hours after the previous day? The answer is that we can't. Through the year, the Sun has sometimes already passed due south by noon, and sometimes it is not there yet, because we measure time by mechanical methods, using devices designed to indicate precise intervals of the time from one day to the next. The reason for this discrepancy between the mean solar day, as measured by Greenwich Mean Time, and the actual behaviour of the Sun is because the Earth moves around the Sun in an elliptical, not a circular orbit, so that its distance from the Sun varies through the year.

The planet also moves at a greater speed around its orbit when it is closest to the Sun, and it is this which affects the time-keeping abilities of the Earth (we must no longer blame the Sun). We will return to this problem in more detail in Chapter eight, when we consider the motion of all the planets. An elliptical orbit is illustrated in Fig 58 on page 150 from which it can be seen that any planet, including the Earth, reaches a point on its orbit at which it is closer to the Sun than at any other time. This point is called *perihelion.* Similarly, there is a point on the orbit at which the planet is at its greatest distance from the Sun called *aphelion.*

The Earth is at perihelion about January 4, when it is some 147 million kilometres from the Sun, and at aphelion about July 4, when it is 152 million kilometres from the Sun (91·8 and 95 million miles

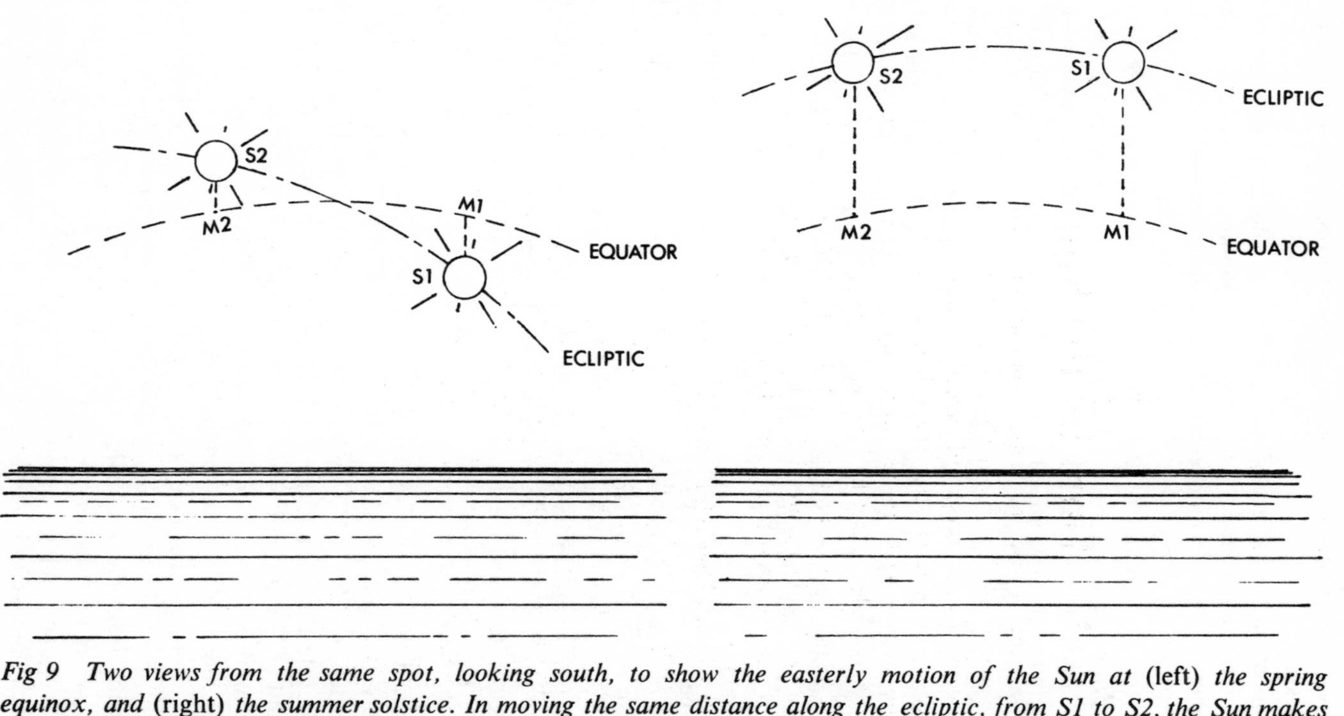

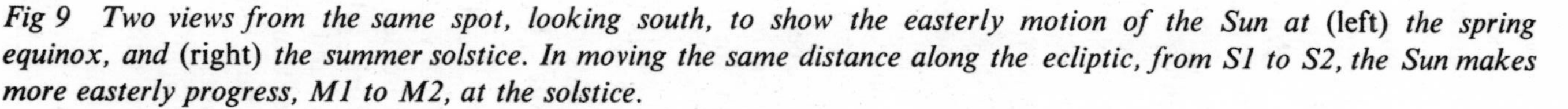

Fig 9 Two views from the same spot, looking south, to show the easterly motion of the Sun at (left) *the spring equinox, and* (right) *the summer solstice. In moving the same distance along the ecliptic, from S1 to S2, the Sun makes more easterly progress, M1 to M2, at the solstice.*

respectively). The solstices, however, occur before these dates by about two weeks.

The most obvious effect of the speeding up and slowing down of the Earth on its path around the Sun is a corresponding effect upon the speed of the Sun in its easterly movement along its apparent path amid the stars. This path, which is our view from Earth of the plane of our orbit seen against the background of the stars, is called the *ecliptic*. The ecliptic can be drawn across star maps because it is virtually fixed in the sky.

The ecliptic circle crosses the celestial equator at the equinoxes, and the angle between them where they cross is $23\frac{1}{2}°$, due to the inclination of the Earth's axis to the plane of its orbit, as shown on the left of Fig 9 so when the Sun is close to the equator, at either of the equinoxes, it is moving across the equator at $23\frac{1}{2}°$. But in June and December, for several days before and after both solstices, the Sun is moving along the ecliptic almost parallel to the equator. As a result, the Sun makes more easterly progress each day near the solstices than at the equinoxes, when it is busily rising above the equator in the Spring, or dropping below it in the Autumn. It is the Sun's east–west progress, in other words its position along the equator, which determines when it will be due south, and hence its timekeeping accuracy. Since the Sun makes slower progress to the east at the equinoxes, it falls behind in its timekeeping. On the other hand, at the solstices it is gaining because it is moving more noticeably in an easterly direction, as shown in Fig 9 on the right.

The Sun's varying eastward movement added to the effect of the Earth's changing speed along its elliptical orbit, produces a curve of 'Sun-fast and Sun-slow' called the *equation of time*. This gives the time which must be added to the local meantime to give the apparent time shown on the sundial. The equation of time is given in Fig 10, and in the table in the appendix.

To avoid the nuisance of having to make corrections to the length of each day through the year, it is assumed that the Sun's path can be represented by a *'mean' Sun* which travels along the celestial equator at a uniform rate throughout the year. This mean Sun gives the mean solar time, and at Greenwich it gives Greenwich Mean Time. At any longitude other than that of Greenwich, the mean Sun will give the local mean time, the time indicated by your sundial corrected by the equation of time.

Internationally, the mean time at Greenwich starting at 00h 00m at midnight, is known as *universal time,* and it is this to which astronomers refer. Universal time is usually abbreviated to UT.

Now we must consider what effect longitude has on local time. The Sun makes one complete revolution of 360° about the Earth in 24 hours. Since each hour is 60 minutes we can work out the correction for one degree

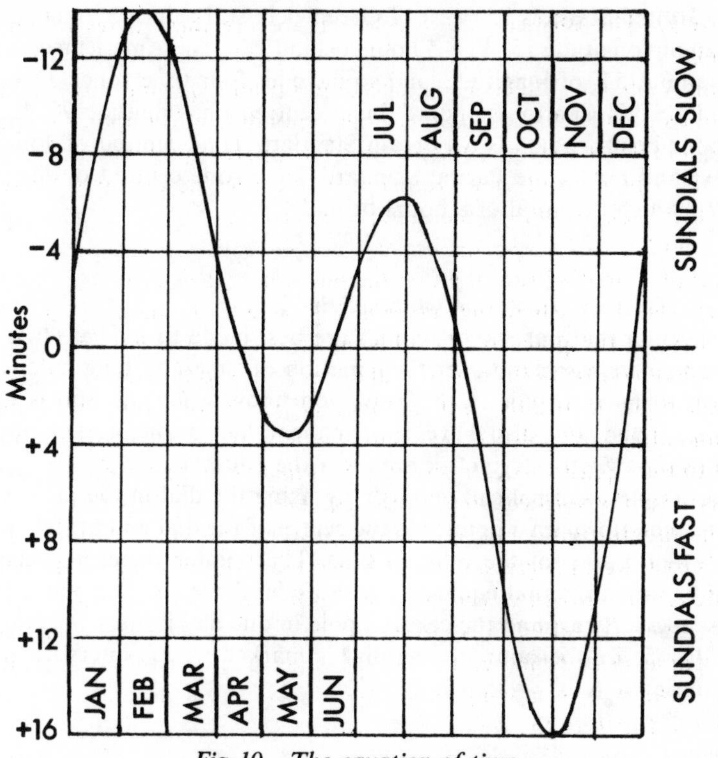

Fig 10 The equation of time.

of longitude by dividing 24×60 by 360, giving 4 minutes. So that for each degree of longitude east or west of Greenwich we must add or subtract 4 minutes from UT to find the local mean time. But do we add or subtract? The Sun, we have seen, rises in the east so that it will be overhead in the east of the country before the west. When it is the mean solar noon at Greenwich the Sun will not have reached due south at a place to the west, so that a sundial in this place will indicate that it is still a few minutes before noon. Accordingly, we must add 4 minutes per degree west to the sundials to find UT, and we must subtract this amount for each degree east of Greenwich.

As the longitude differences become considerable, the mean standard time (which is always based on the time the 'mean sun' is due south) has to be advanced on UT if the place is to the east of Greenwich, or delayed if to the west. For each standard mean time zone on the Earth, the clocks are altered one hour relative to the adjacent zone. This represents 15° of the Sun's movement, and hence the centre of each time zone has a longitude 15° east or west of the adjacent ones. Thus, the eastern seaboard of the United States is 75° west of Greenwich, so Eastern Standard Time, centred on longitude 75°W, is 5 hours behind UT. The American continent spans about 55° of longitude, and so includes four time zones. One hour behind EST is Central Standard Time centred on longitude 90°W, then one hour behind again is Mountain Standard Time centred on longitude 105°W, and finally the Pacific Standard Time zone centred on longitude 120°W, which is a total of 8 hours behind UT.

Sundials

For astronomical purposes, sundials are best made to indicate true local solar time: the correction for the equation of time and longitude is easy enough to make. Made in this way, you know when the Sun is on the meridian—and you also have a permanent indication of the southerly direction (or north, if you live south of the equator).

The simplest sundial can be made by fixing the dial in the plane of the celestial equator with a pointer in the centre of the dial and at right angles to it which points at the celestial pole. The pointer therefore makes an angle to the horizontal equal to the latitude. Since the Sun moves more or less regularly around the celestial pole in one day it must turn through 360/24=15° in one hour, so the dial is marked in 15° intervals around the pointer.

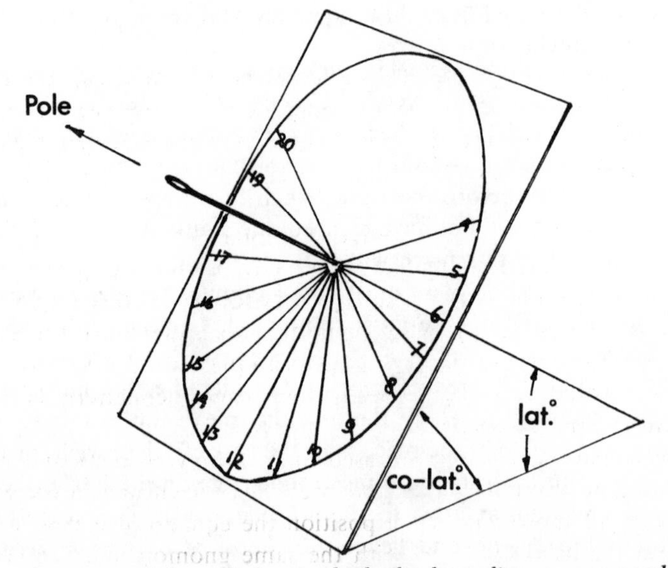

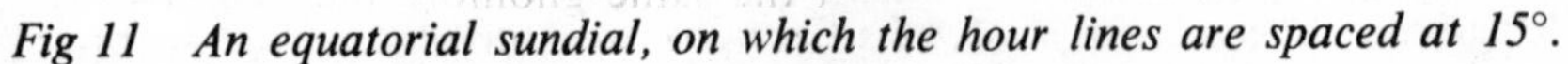
Fig 11 An equatorial sundial, on which the hour lines are spaced at 15°.

The pointer is called the *gnomon* or *style*. The two terms have come to mean virtually the same thing, although 'style' usually refers to the familiar triangular gnomon seen on the most common horizontal dial as used on ordinary garden sundials. This simple form of sundial with a gnomon at right angles to the dial is called the equatorial dial, but it is of limited use since the Sun can only cast a shadow on the dial plate when it is on your side of the celestial equator.

An equatorial sundial is so simple to make that it is worth the small effort. Draw a circle of diameter 15cm or more on a plain card, and draw diameters at 15° intervals, measured with a protractor. Mark the right hand end of one diameter 6h, then moving clockwise, mark the end of the next diameter 7h, and so on round to 18h. All equatorial (and horizontal) dials for the southern hemisphere must be numbered anti-clockwise. A long needle in the centre of the dial and at right angles to it will make a satisfactory gnomon. Cut two pieces of card to support the dial at an angle equal to that of the equator at your latitude. This is equal to the co-latitude, or (90-lat)° as shown in Fig. 11. The supports can be glued, or fixed in position with sticky tape. Note that 12h is the lowest point on the dial.

Check with a spirit level that the model is positioned horizontally, and turn it until the shadow of the needle falls on the true local solar time—that is, the Universal Time plus the corrections for the equation of time and the longitude of the site. A more approximate alternative would be to line up the gnomon on the pole star and leave the dial in position until the following day.

A very interesting result can be obtained if you set up the dial a few days before the autumn equinox using the Sun to set the dial accurately. Then observe how the shadow disappears some days later as the Sun drops below the plane of the dial.

HORIZONTAL DIALS

To be able to use the sundial all the year round, the plane of the dial must be horizontal, which results in a distortion of the simple symmetry of the equatorial dial. If we imagine that we could leave the gnomon in the same position but turn the equatorial plate until it is horizontal, the gnomon would cast the same shadows, but the pattern of successive shadows would be changed.

The equatorial dial can be used quite simply to mark out a dial for mounting at any other angle, but we will concentrate on the horizontal dial. Fig. 12 shows that if we position the equatorial dial so that it just rests on the horizontal dial with the same gnomon, the 12h line on the equatorial dial will touch the horizontal dial, so this will be the 12h line

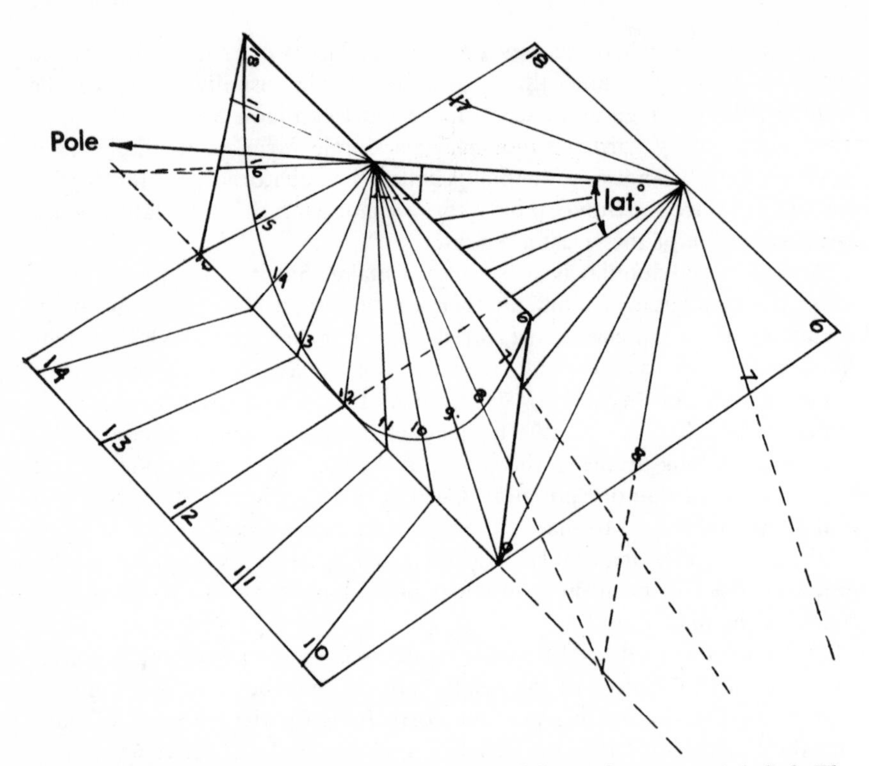

Fig 12 The horizontal dial can be projected from the equatorial dial. The equatorial dial fits between the horizontal dial plate and its style.

for the horizontal dial also. The other lines can be projected through the equatorial dial on to the horizontal surface, as if the equatorial dial were transparent.

We can see that the lines projected from the equatorial dial meet the lines on the horizontal dial. In addition Fig 12 shows that the style meets the horizontal dial at an angle equal to the latitude. The equatorial dial is at right angles to the style and is of a radius such that it fits between the style and the plate. We now have all the information we need to construct a horizontal dial. Fig 13 shows the construction of a horizontal dial for latitude 51°. Draw the meridian NS, and at S draw a line at the angle of your latitude, which is 51° in this example. Draw the line EW at right angles to NS and at a distance from S according to how large you want your sundial to be. EW crosses NS at B.

From the style (the line SA at 51° to NS) draw AB at right angles to meet EW at B. Note that the angle ABS is equal to your co-latitude.

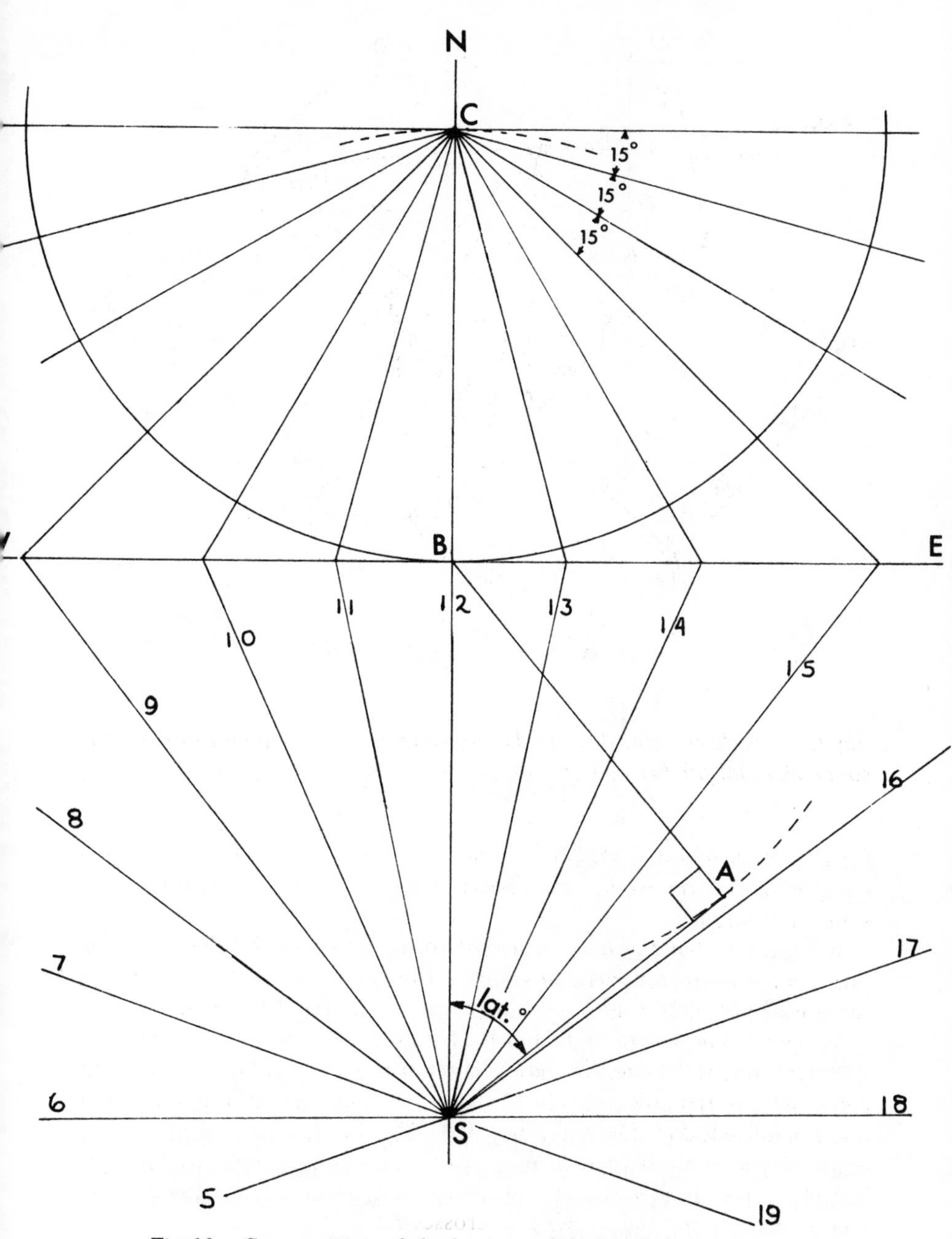

Fig 13 Construction of the horizontal dial and the style.

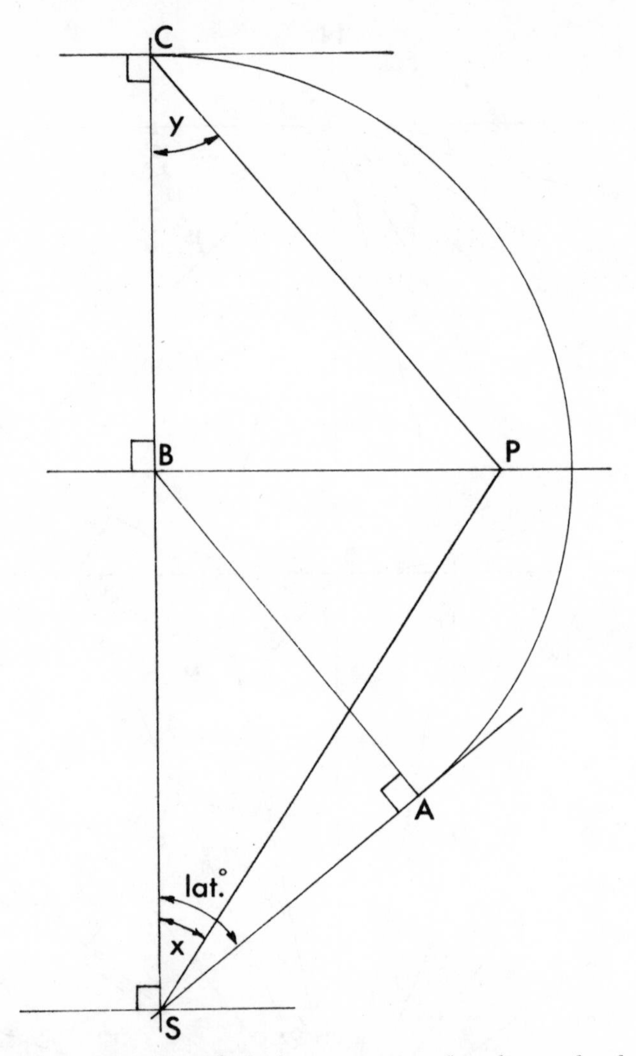

Fig 14 Construction to find the expression for the angles between the hour lines and the meridian on the horizontal dial.

Set your compasses at radius AB, and with the centre at B draw an arc on NS, closer to N than S. This arc crosses NS at C. Keeping the same radius, draw a circle with centre C, which will, of course, just meet EW at B. This is the equatorial dial.

Draw a diameter of the equatorial dial at right angles to NS (ie through C) then with a protractor mark angles at 15° intervals starting from this diameter around the half of the circle touching the horizontal dial. From the centre C draw lines to meet the line EW. Note that the lines at 15° to the east–west diameter of the equatorial dial cut EW only when projected a long way. If these prove to be too difficult to draw in this manner we can get round the problem by calculating the required angles at S as described below. The hour lines can now be drawn on the horizontal dial from S to each of the points on EW where the hour lines of the equatorial dial intersect. The 6h–18h line on the horizontal dial will be at right angles to NS through S.

Fig 14 shows how we can calculate the required angles at S, which, although involving more mathematics, avoids the difficulties of the drawing method. Fig 14 contains the essential construction lines for one of the hour lines on the horizontal dial from the corresponding line on the equatorial dial. The equatorial dial is at centre C, as in Fig 13, and an hour line CP has been drawn at an angle y to the meridian CBS. The line on the horizontal dial will then be SP, which is at an angle x to the meridian CBS. An arc of the circle at centre B, has been drawn from the end of the style at A to C, so that the length of the line BC must be the same as that of the line BA, since they are both radii of the same circle.

We see that:

$$\text{Tan}\, y = \frac{BP}{CB} \text{ but } CB = BA, \text{ so } \text{Tan}\, y = \frac{BP}{BA} \quad (1)$$

$$\text{Tan}\, x = \frac{BP}{SB} \quad (2)$$

$$\text{and Sin(lat)}^\circ = \frac{BA}{SB}$$

$$\text{so that } SB = \frac{BA}{\text{Sin(lat)}}$$

Substituting in equation (2) we get

$$\text{Tan}\, x = \frac{BP\ \text{Sin(lat)}}{BA}$$

and since, from equation (1) we know that $\frac{BP}{BA} = \text{Tan}\, y$, we have

$$\text{Tan}\, x = \text{Tan}\, y\ \text{Sin(lat)}$$

Since the sine of the latitude will be constant for any particular dial, we can now substitute values of 15°, 30°, 45°, etc for y and hence find the corresponding values of x, the angle on the horizontal dial. For example, at latitude 50°, the hour line for 10h on the horizontal dial will be given by Tan x=Tan 30 Sin 50, so the 10h hour line will be at an angle of 23·8° to the 12h line (and so will the 14h line). The style for the dial will be the triangle SAB in either Fig 13 or Fig 14. Only the side SA and the angle

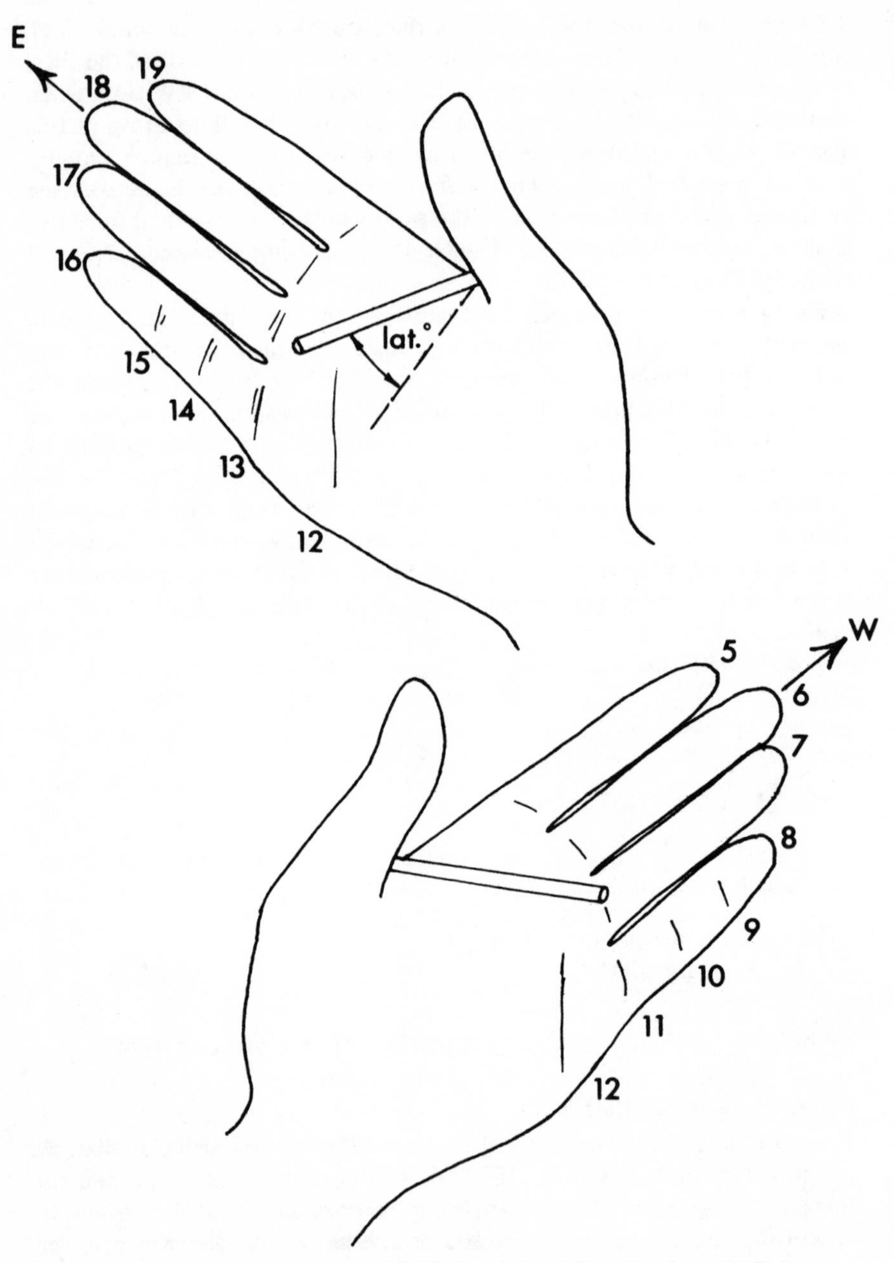

Fig 15 How to use your hands as approximate horizontal dials.

BSA are critical, and you can make the side AB curved, if you wish, to add a bit of decoration to the dial.

The dial can be drawn on card, with a thin card style mounted at right angles to the dial with its base on the 'sub style' line BS, taking care to get the angle equal to your latitude to just touch the point S.

If you are a little more ambitious, you can try a more permanent monument for your garden, and make the dial and style in brass. Use thin engraving quality brass, about 2·5mm would do, and scribe the lines with a sharp, hard point on the dial using marks transferred from a pencil drawing on paper. You must allow for the thickness of the style in the layout of the dial, however, which is ignored in the case of a thin card style. When you have drawn the plan as in Fig 13 cut the plan carefully in half along the line BS, and separate the halves by the thickness of the style. The finished dial will then look very like Fig 13, except that the line SB will be a thick line.

Mounting the style on a brass dial can present some problems, and the handyman with some ingenuity may think of his own particular solution. One possibility is to drill a series of holes along the dial at the base of the style, using a drill a little smaller in diameter than the thickness of the style. Then with a slim file you can file a straight-sided slot through which the style can be pushed to the required height. It can be secured in position on the underside of the dial with a modern adhesive of the epoxy resin type. Make sure that the style is held vertically to the dial while the adhesive sets.

A 'HANDY' DIAL

By a strange coincidence, the geometry of the horizontal dial for the intermediate latitudes on Earth corresponds approximately with the average proportions of the human hand, and you can use this fact to make an 'instant' sundial at any time using your hand as the dial plate, and any straight, fairly thin object, such as a pencil, as the style.

In the Northern hemisphere imagine that the morning hours are written around your left hand and the afternoon hours around your right hand, as shown in Fig 15. (In the Southern hemisphere it will be the morning hours on the right hand and afternoon hours on the left.) In both cases, the 12h mark would be beneath the joint of the little finger on the edge of the hand. The next mark, 11h on the left hand or 13h on the right hand, will be at the base of the little finger, the next mark at first joint of the little finger, the next at the second joint, the next at the tip of the little finger, and the remainder at the tips of the other fingers.

The style is held by the thumb against the side of the hand so that it makes an angle with the palm, as close as you can judge, equal to your latitude. Using the 'morning' hand, point the '6h mark', which is the middle finger, to the west, and now read the sundial time by the position of the shadow on your hand. Using the 'afternoon' hand, you must point it due east.

This little novelty may be used to tell the time, if you know where the east and west are or, knowing the time, you can find due east and west by turning your hand until the shadow reaches the right time. You must hold the palm upwards and horizontal, of course.

OTHER TYPES OF DIAL

There is a fantastic range of different types of sundial, enough to fill a large book. For our experimental approach to astronomy, there are two types of dial which are of particular interest. The first of these is an *analemmatic* dial; this has a moveable gnomon and is of particular interest to us because it can be combined with the conventional horizontal

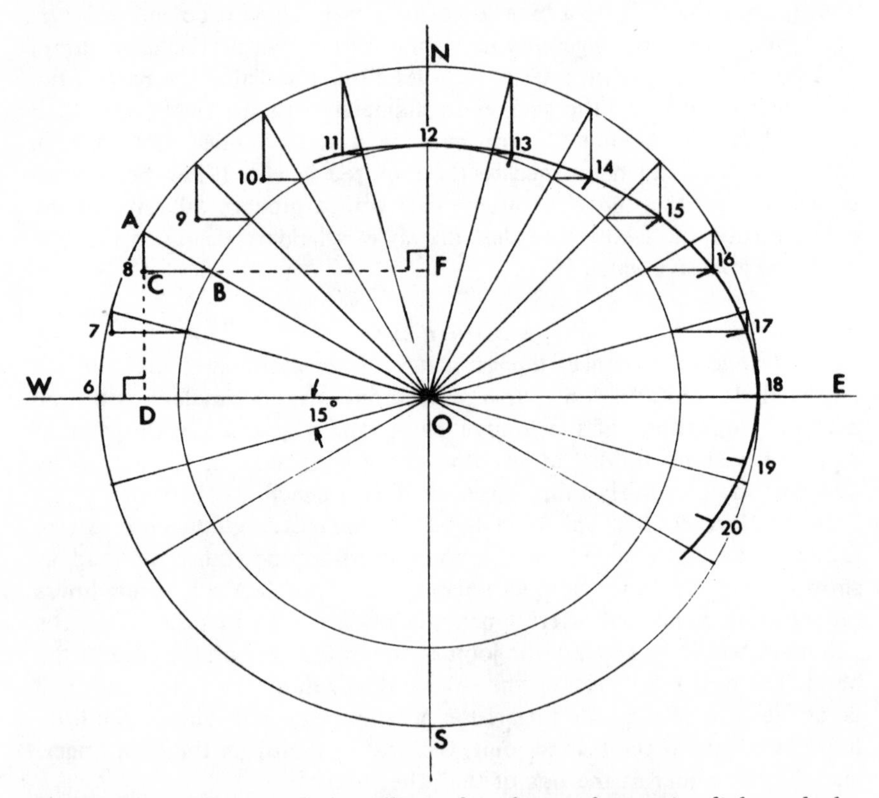

Fig 16 Construction of the ellipse for the analemmatic dial, and the position of the hour lines on the dial, for a latitude of 51°.

dial to make a completely portable unit, which also automatically indicates north and south.

The construction of the analemmatic dial is rather more complicated than the horizontal dial. The theoretical basis for the construction is also much more complex, so we will just content ourselves with a step-by-step description of how to make this type of sundial.

This dial has an elliptical scale; the gnomon is a vertical rod which is moved along a line on the meridian, the exact position depending upon the time of the year. We will draw an analemmatic dial with a conventional horizontal dial on the same axis on a piece of paper which later will be stuck on to a board. A sewing needle will be used as the moveable gnomon of the analemmatic dial and the style for the conventional dial can be cut out of thin card.

The dial is an ellipse, the major axis of which can be as large as you like to suit the paper available. The minor axis is the 'major' axis multiplied by the sine of the dial's latitude. Lightly draw two lines NS and WE crossing at O at right angles, as shown in Fig 16, and draw the outer circle at any radius from O. On the same centre O, draw a second circle of diameter equal to the larger circle times sine (lat)°.

Now draw radii to the outer circumference at 15° intervals, such as the radius OA, starting and finishing 30° below WE. At each point where these radii cross the inner circle, such as at B, draw a line between the two circles at right angles to the vertical line NS, such as the line CBF. Where the radii meet the outer circle, such as at A, drop a vertical line to meet WE at right angles, such as at D. Where the two construction lines from each radius intersect, such as at C, gives the point for one of the hour marks on the elliptical scale. For example, C is the mark for 8h 0m. The construction is shown in some detail in Fig 16, and the ellipse itself is drawn in the right hand half. A very versatile drawing instrument for completing curves such as this without having to plot an enormous number of points is a flexible plastic-covered metal strip sold for draughtsmen.

Having completed the dial marking, rub out the construction lines, except for the lines WE and NS. The construction of the scale for the gnomon is rather tedious and involves numerous construction lines which must be erased afterwards, so draw them lightly. To avoid confusion, this construction is shown separately in Fig 17 for one of the gnomon positions only. If you look at the completed analemmatic dial at the bottom of Fig 18 you can see what the finished scale for the gnomon will look like.

The positions for the equinoxes lie on the line WE at O, the centre of the two circles. The remaining dates are arranged symmetrically about

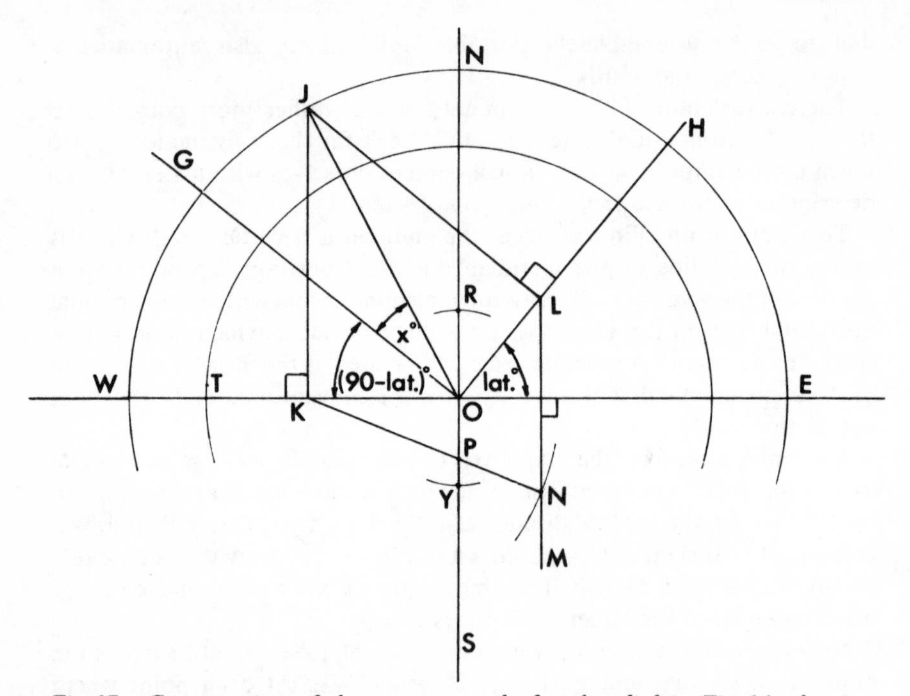

Fig 17 Construction of the gnomon scale for the dial in Fig 16, showing how the positions R and Y are found for one value of the Sun's Dec, x°.

WE, along the meridian NS. It is assumed that the Sun reaches equal midday altitudes on each of the pairs of dates marked on the scale, and although this is not strictly true, it is close enough for our purpose.

In the construction of this and the last sundial to be described in this chapter we need to know the angular distance of the Sun above or below the equator at the various dates marked on the scale. This distance is known as the Sun's *declination,* one of the two main co-ordinates used to measure the position of astronomical objects on the celestial sphere, and we shall examine these co-ordinates in more detail in the next chapter. The values of the Sun's declination are given in the following table (see also the Appendix):

Dates:	Sept 22/Mar 22	Aug 22/Apr 22	Jul 22/May 22	June 22
Sun's declination:	$+0°$	$+12°$	$+20°$	$+23\frac{1}{2}°$

During the winter, the declinations are almost the same for the corresponding dates after the autumn equinox, except that they are prefixed with a minus sign to show that they are below the equator. The Sun's declination for the winter dates can be found in a table in appendix B.

Returning to Fig 17, draw the line OH at an angle to OE equal to your latitude. Then draw OG at an angle to WO equal to your colatitude (which is 90° minus your latitude). The same thing can be achieved by drawing OG at right angles to OH. Now draw a radius to the outer circle OJ at an angle $x°$ to OG equal to the declination of the Sun on the date in question, taken from the table. From J, draw a line to meet OH at right angles, and another to OW, also meeting it at a right angle, to give

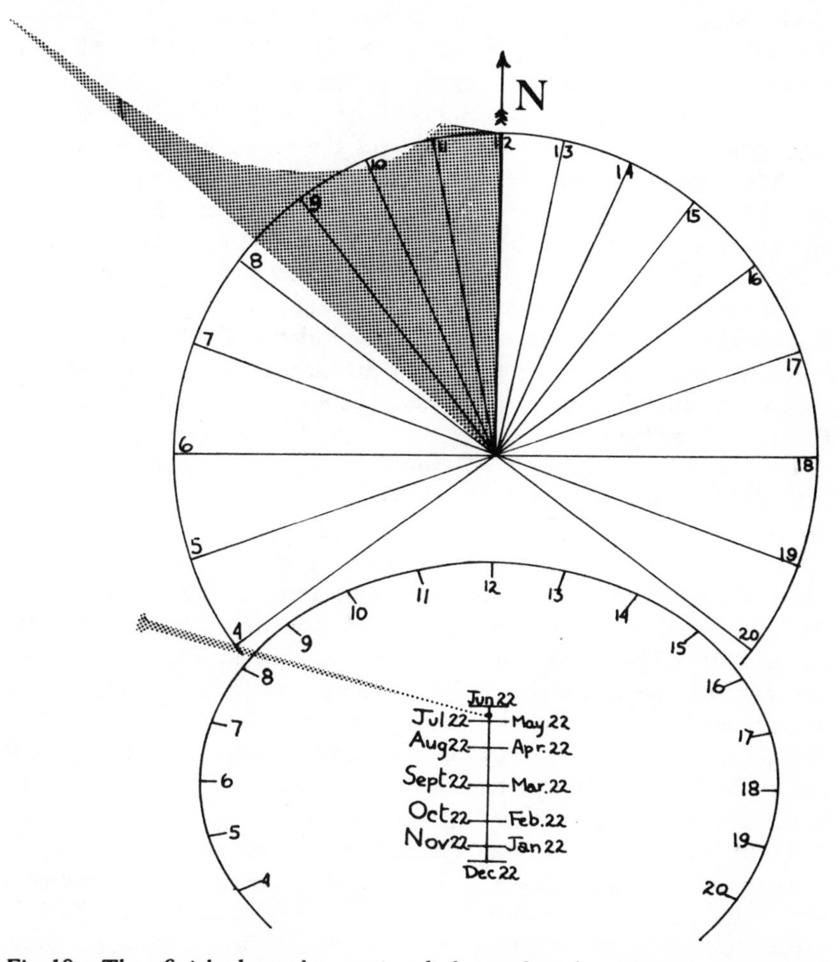

Fig 18 The finished analemmatic dial combined with a conventional horizontal dial, indicating 8h 15m on June 1 or July 1 (or close to these dates).

the points L and K respectively. From L, drop a line perpendicularly through OE to somewhere below EW at M. Now you must find the point N on LM which is at the same distance from K as the radius of the inner circle. So set your compasses to the distance OT, and with the centre at K strike an arc of this radius across LM to give N. Draw the line KN, which crosses OS at P. You may not believe this after this long rigmarole, but the length PN is the length you want above and below O on NS for your gnomon scale; so set the compasses to length PN and with centre O strike the arcs across NS at R and Y.

You must repeat this procedure for each of the pairs of dates given in the table above for declinations of 12°, 20°, and 23½°, and for any intermediate dates you may think necessary. The date scale for a dial for a southern latitude must be the reverse of that shown in Fig 18. When you have erased all the construction lines for the analemmatic dial, redraw the north–south line lightly so that you can draw a conventional horizontal dial on the same line, as described at the beginning of this chapter, and as shown in Fig 18. Glue the combined dials on to a board and fix a conventional triangular card style on the horizontal dial. Push a long needle into the scale on the analemmatic dial at the date and, holding the dials horizontal, turn the board until the shadows of both gnomons fall on exactly the same time on their respective scales. This is the local solar time which must be corrected for the equation of time and longitude to give the UT or the mean time for your zone. In addition, the centre line of the dials will be aligned north–south, the north being in the direction of the 12 hour lines on the dials.

THE CAPUCHIN DIAL

The dials described so far indicate the time according to the azimuth of the Sun, whereas the 'Capuchin' sundial measures the time from the Sun's altitude. Now the Sun's noon altitude, or at any time of day, changes throughout the year as its declination changes, so that, like the analemmatic sundial, any dial which operates on the altitude of the Sun will have to involve an indicator adjustable for the time of year. The theory behind the construction will not be described, but to a great extent the principle is self-evident; the dial measures the altitude of the Sun with a plumb bob, and the plumb line is set to suit the date.

The finished dial is shown in Fig 19 together with some of the constructional lines and references, which you should erase from the finished article. Use a rectangle of fairly thick artist's card, measuring about 15cm by 20cm. You can make it smaller if you want a really pocket-sized dial because, like the combined horizontal and analemmatic dials, the Capuchin is completely portable.

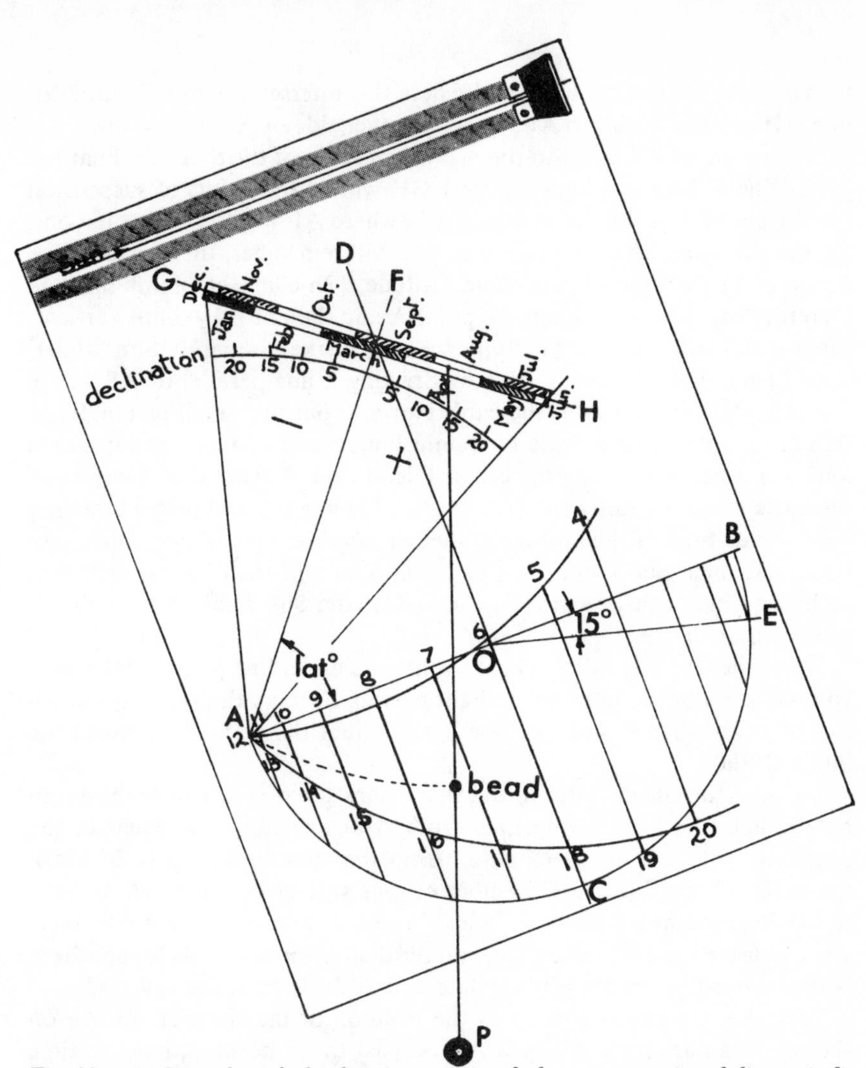

Fig 19 A Capuchin dial, showing some of the constructional lines, indicating 7h 20m or 16h 40m on August 20 or April 15.

Draw a vertical line CD down the centre line of the card. Then, at about one-third of the distance along this line from the bottom, draw line AB at right angles to CD, crossing it at O. Now with centre O draw a semicircle to nearly fill the space below AB. Divide the circumference ACB into 12 equal parts by lightly drawing construction lines at 15° intervals from the centre O, such as the line OE, and from each point on the circumference drop a line perpendicular to AB.

From point A, which is the 12h position, draw the line AF at an angle to AB equal to your latitude, and where this intersects with CD draw the line GH at right angles to AF. Now on either side of AF draw a line from A at an angle to AF equal to the maximum value of the Sun's declination, 23½°. Where these two lines intersect GH will be the points of suspension for the plumb line for the solstices, and where AF meets GH is the point for the equinoxes. At the equinoxes, you will remember, the Sun is on the equator. At the time of maximum altitude, 12h 00m, the plumb line will therefore hang down through the point A and the line AF will be vertical; the line AB will point to the Sun. We must make a separate Sun 'finder' clear of the dial, and we do this by drawing a line parallel to AB above the date scale at the top of the card. Now cut out two small rectangles of thin card, about 14mm wide by 26mm long, and cut a slot about 20mm long and 1mm wide along the centre line of each. Ensure that the sides of this little slot are cleanly cut and parallel. Placing the two pieces together, bend about 4mm of the bottom of the two 'legs' on each in opposite directions and then glue them back to back. You will then have a tall, thin archway which can be glued to the end of the Sun finder line above B, as shown in Fig 19.

When the Capuchin dial is held so that the finder line points to the Sun, you will see a bright band of sunlight passing through the archway, which can be easily aligned with the line itself so that the dial is inclined at the Sun's altitude.

To cater for all the dates in the year, mark the first day in each month on the line GH by drawing lines from A at angles to AF equal to the respective values of the declination from the table in appendix B. Mark the months from June to December on one side of the line GH, and the remaining months on the other side. If you like, as a permanent reference, you can leave the declination scale on the dial. Capuchin dials for southern latitudes must be made with the date and declination scales reversed.

With the compasses centred at the position of the summer solstice on GH, draw an arc from A below AB. Repeat this procedure, but this time with the centre at the winter solstice, and draw the arc which crosses AB, to give the 'Capuchin-hood' shaped scale. If you wish you can add more vertical lines to indicate the half and quarter hours by suitably subdividing the circumference of the semicircle. The hours of the day can then be marked, the morning hours along the top arc to 12h at the point A, and the afternoon hours (using the 24 hour clock times, of course) on the lower arc.

With a very sharp knife or a safety razor blade, and using a steel rule as the guide, cut along the line GH between the dates of the solstices. A piece of thread long enough to hang from the slot you have just cut to about 12cm below the card must now be passed through the slot. The thread will be held firmly in the slot even with some slack behind the card. One end of the thread is fastened with a little glue behind the card. On to the other end in front of the dial slip a small bead. If it is a good fit on the thread the bead will stay where you put it. If it fits loosely it should be glued to the thread so that it hangs in front of the scale while there is still plenty of slack thread behind the card. Finally attach a small weight to the end of the thread; a steel nut is ideal.

To use the Capuchin dial, pull the thread along the slot so that it emerges at the present date. Holding the card upright so that the thread hangs alongside the dial, tilt the Sun sighting line upwards until the plumb line passes through the 12h mark, A. Now either move the bead on the thread, or pull the right amount of thread through, so that the bead is on point A when the thread is hanging freely. The dial is now set up for the day.

Remembering to keep the face of the dial vertical so that the bead hangs just clear of the dial, point the Sun finder at the Sun so that the thin beam through the arch is bisected by the line. As always with experiments in which you must make a Sun sighting, DO NOT look through the sights at the Sun itself; it could damage your eyes. This is why the bright line sight is so convenient; you can watch the card while aiming it at the Sun over your shoulder.

The time is given by the vertical line on the scale where the bead touches it. In Fig 19 for instance, the dial is indicating 7h 20m, or is it 16h 40m? If you cannot decide which it is, as might happen close to noon, you must make another sighting a little later. If the Sun's altitude has reduced since the first sight, then it is in the afternoon. If it hasn't, it isn't!

The Capuchin dial is an interesting little instrument and demonstrates several facts about the Sun throughout the year. If you set the plumb line to the summer solstice position, then tilt the dial until it reads 12h 0m, you can see the maximum altitude attained by the Sun in the year: it is the inclination of the Sun-sight line. Similarly you can see just how low the Sun is at midday on the winter solstice.

If you keep the declination scale used in the construction of the dial, you can read the declination of the Sun for any date by setting the plumb line to that date, and inclining the dial to the 12h 0m position. The declination is read where the line crosses this scale.

By setting the plumb line to any date and tilting the dial until the Sun-sight line is horizontal (which can be done quite easily by checking that the plumb line is parallel to the vertical lines marking the hours on the time scale), you can read the sunrise and sunset times for that date where the plumb line crosses the time scale. Remember to correct all readings for the equation of time.

Experiments with sundials

Sundials can be used as instruments to measure other than local mean solar time. We have already seen how the analemmatic and horizontal dials combined can be used to find due north and south, and how the Capuchin dial indicates sunrise and sunset times and so on. By comparing sundial time with a clock showing mean time, you could therefore estimate your longitude using one of your portable sundials, provided that you are at the latitude for which the dial was made.

You can use any sundial to measure the value of the equation of time on any day of the year. Simply compare the time by your watch (always assuming it keeps very good time); better still, use a stop watch set by one of the daily time signals and compare with the sundial when it indicates noon.

Since the Moon follows the Sun's path across the sky fairly closely, we can use the Moon shadow on our dial to tell the time at night. At the first quarter, for example, the Moon sets after the Sun. It is to the east of the Sun and one quarter of the way around the ecliptic. On our conventional horizontal dial, therefore, we can regard the Moon as the Sun, running one quarter of a day slow. The time the moonlight shadow indicates on the dial will then be 6 hours slow. At full Moon it will be 12 hours slow, at the last quarter 18 hours slow—or 6 hours fast, which comes to the same thing—and so on. In fact since the Moon takes a lunar month or 28 days to make one trip around the ecliptic, it is, very roughly, 1 hour slow for each day past new moon. This assumes the lunar month to be 24 days, the same as the number of hours in a day.

One final word on setting up a permanent horizontal dial in your garden. Obviously you should try to situate it where it will be in sunlight for as much of the day as possible, but other considerations frequently overrule this ideal—the sunniest spot in the garden might be in the middle of the goldfish pond. Having decided upon its exact spot, set the dial on a temporary support while you line it up on the meridian. Do this using the sundial time calculated for local noon, working from 12h 00m UT, corrected for your longitude, and for the equation of time. Let's say that this time comes to 11h 49m. This is the time your sundial must read at noon UT on that day. Knowing the exact position of the dial, you can erect the base to the same orientation, and the dial will then sit squarely on its base.

EXERCISES ON CHAPTER 2

1 On a ship crossing the Atlantic on the latitude of Greenwich we make a Sun sighting with a Capuchin sundial, which indicates a time of 14h 25m. According to our chronometer, which is set to UT, the time is 15h 02m. What is our approximate longitude?

2 We wish to set up a conventional horizontal sundial in the garden, but find that the Sun is invisible due to cloud at noon and for two or three hours afterwards. If the date is March 22, how can we align the dial without using any other instrument, even a clock?
3 When could you use the horizontal dial to tell the time by the Moon? Could you use the analemmatic or Capuchin dials as a Moon dial?
4 Having completed a horizontal dial, of the type shown in Fig 13 you take it outside to set it up to read the time. The longitude of your garden is 6° west of the centre of your mean time zone, the date is October 23 and the time by your watch is 10h 48m. To what time should you set the sundial?

3
Time by the Stars

What angle does the Earth turn through in a day? A full circle of 360°? This is what we assumed in the last chapter, but if we examine the true state of affairs we find a problem—the Sun is crossing the sky towards the east each day in its yearly journey around the celestial sphere.

As we saw in chapter one, the stars are so far away that they may as well be regarded as fixed on the inside of a sphere of an enormous radius. So enormous, in fact, that even on the Earth, swinging round the Sun at a distance of 150 million kilometres we cannot detect any change of our apparently central position in the celestial sphere by parallax.

As well as the stars, we can imagine that the celestial equator, the ecliptic, the poles, and the celestial 'grid reference' system of *right ascension* and *declination* are permanently fixed on this sphere, and that the Sun, Moon, planets, comets, meteorites, artificial satellites and any other heavenly bodies belonging to our own solar system that you can think of, all move across this unchanging background.

Fig 20 shows the celestial sphere as seen from the outside, with the Earth at the centre. If we plot all the points which are, say, 40° north of the celestial equator we will obtain a circle parallel to the equator at the position shown. We can plot such parallels at any angle we like, both north and south of the celestial equator, and the resulting circles will correspond exactly to terrestrial latitude. This is the *declination* of the object (abbreviated to Dec, or δ).

The declination of an object is the angle it makes north or south of the celestial equator, and is measured in degrees. Declination north of the celestial equator is denoted with a + sign, and south of the equator with a − sign. Since declination corresponds exactly with terrestrial latitude, the declination overhead anywhere on Earth equals the latitude of the place.

The other co-ordinate of a star's position is *right ascension* (RA, or α), which corresponds to terrestrial longitude, but instead of being measured eastwards and westwards from the zero line, right ascension is measured eastwards along the celestial equator from the true equinox.

Any star is on a circle passing through both celestial poles. This is its

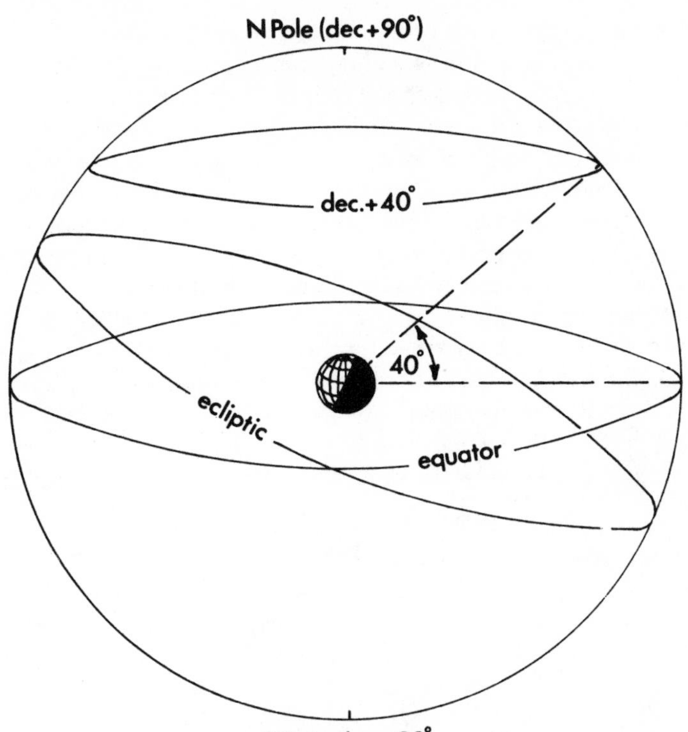

Fig 20 The Earth, in a very much reduced diameter celestial sphere, showing how a parallel of Dec is projected at an angle to the plane of the equator.

hour circle of RA. The RA is measured as the interval of time between the transit of the equinox on the observer's meridian and the transit of the hour circle of RA of the star. The hour circle through the equinox, therefore, is RA zero hours. Fig 21 shows a number of hour circles of RA on the celestial sphere. The equinox (the First Point of Aries), is shown by the symbol ♈. Six hours after the transit of the equinox, the RA on the meridian will be 6h00m. Stars with RA 9h00m will still be to the east of the meridian. Stars with RA 3h33m, for example, will have already crossed the meridian, and so on.

So now we have two frames of reference; the altitude and azimuth of the objects in the sky, which is a frame of reference fixed relative to the observer on the Earth; and the RA and Dec, which is a frame fixed relative to the celestial sphere. These two frames of reference are constantly changing in their positions relative to one another, and are related by time.

As the Earth turns once on its axis completing a day of 24 hours of mean solar time, the Sun moves along the ecliptic towards the east. How much is this daily movement on the ecliptic? The Sun must travel through 360° on the ecliptic in 1 year of 365¼ days. This is approximately 1° each day. Although this may not seem very much, it is about twice the angle the Sun subtends to our eyes, or, as we found in the last chapter, it is 4 minutes of time. This means that the Sun reaches the meridian 4 minutes later than the point it occupied in the sky on the previous day. It is later because the Sun's movement on the ecliptic is eastwards.

The phenomenon can be shown as in Fig 22 in which the Earth's position is shown on two dates fairly close together. The Earth has turned on its axis through 360° (several times) as is measured by the transit of the same star across the meridian. In the first case the Sun is also on the meridian, but in the second case it has not yet reached the meridian.

If time were regulated by the stars, each 'day' would be about 4 minutes

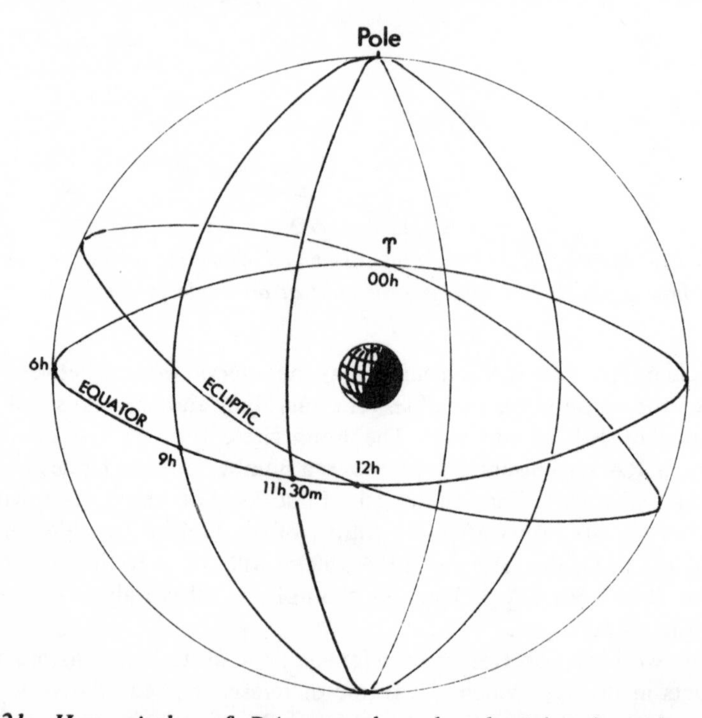

Fig 21 Hour circles of RA pass through celestial pole and cross the equator at right angles. Each hour circle is measured as the time interval of its transit since the transit of the First Point of Aries from which it is measured eastwards.

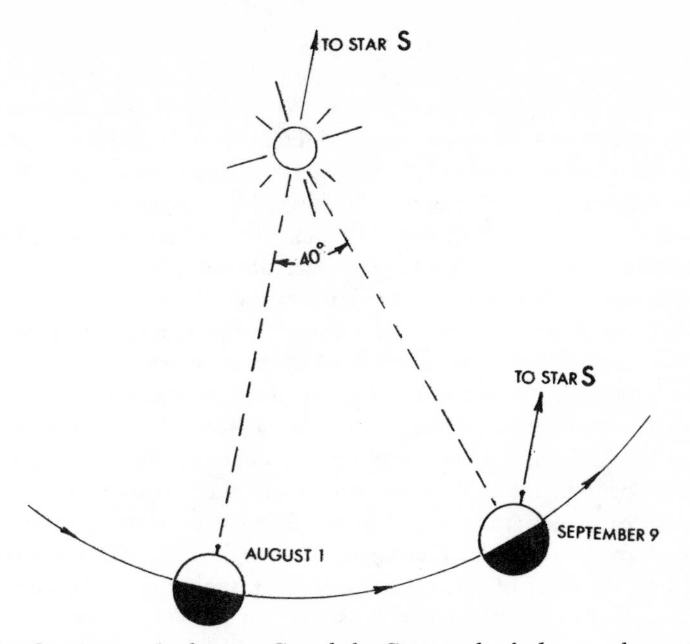

Fig 22 On August 1, the star S and the Sun are both due south at midday. On September 9, however, star S is due south before the Sun, so that time by the stars gains on solar time.

shorter than the day as we know it. At midday, the same star would be on the meridian, and all the other stars would be in the same position in the sky because the Earth would have turned exactly through 360°. Even though it was 'midday', you might well be able to see the stars because the Sun might be on the other side of the sky. Being pitch dark at midday would be a little inconvenient, so we do not use *sidereal time*, time by the stars, except in astronomy.

Since the stars are fixed in RA and Dec, the hours we use for measuring RA are sidereal hours, although the difference of only 4 minutes in 24 hours is difficult to notice during a short period (it is only about 10 seconds per hour).

Sidereal time is the interval to have elapsed since the transit on the meridian of the First Point of Aries. It is therefore related to the observer's longitude on Earth, just as is the local mean time. When the true equinox is on the meridian, the sidereal time is 00h00m. It follows from this that the right ascension of any object on the meridian is equal to the local sidereal time.

The first Point of Aries is in transit on the meridian at Greenwich at 00h00m UT close to the autumn equinox, so at midnight on the autumn equinox the sidereal time is the same as Universal Time. From then on, the sidereal time gains about 4 minutes, or 3·9 minutes to be more precise, on UT every 24 hours of mean time. At midnight, two days after the autumn equinox, a clock showing sidereal time will have gained 7·8 minutes on a clock showing UT, and so on through the year until, by the next autumn equinox, the sidereal clock will have gained a whole day, and will once more line up with the UT clock. If the hour circles of RA were drawn upon the sky we could tell the sidereal time at any moment, simply by seeing which one was due south, on our meridian.

We can easily measure the interval which has elapsed since an object was on our meridian to the south, and in many ways this is more useful than the object's RA. This interval is called the *hour angle* (HA).

The hour angle of an object is the interval since the object was on the observer's meridian, and is therefore the angle to the object measured westwards around the celestial equator from the meridian, and like RA, is usually expressed in units of time—in sidereal hours, minutes and seconds. It follows that the hour angle of an object is the difference between its RA and the hour of RA on the meridian, the latter being the sidereal time. The hour angle of a star is illustrated in Fig 46 (p.119).

We can also see that since the sidereal time is the interval which has elapsed since the transit of the First Point of Aries, the hour angle of the First Point of Aries is the sidereal time.

You can now see how all these various quantities begin to tie together. We can summarise them as follows, remembering that hour angles are measured westwards from the meridian, and RA is measured eastwards from the equinox. The hour angle = local sidereal time − RA, for any object.

If, as is sometimes more convenient, we wish to express the hour angle of an object to the east of the meridian in terms of the smallest angle, for example 2 hours instead of 22 hours, we must prefix the angle with a minus sign so that the relationships explained above remain true. As an example, we will assume you have calculated the sidereal time as 5h 38m, using the methods which are explained shortly. What is the hour angle of a star at RA 11h 40m, and where will it be in the sky if it has a Dec of about 0°? HA = ST − RA, and so in this case HA = 5h 38m − 11h 40m. This gives an hour angle of − 6h 02m, which is the same as 17h 58m. The star is to the east of the meridian, and since it is close to the celestial equator at Dec 0° and about 90° from due south, the star is almost due east of the observer and is very near the horizon.

Calculating sidereal time

How do we find the sidereal time for any moment in the year? At the autumn equinox the sidereal time is the same as our local mean time, that is, UT corrected for our longitude. One year later, mean and sidereal time will once more coincide. If a sidereal time clock gains 24 hours in a year, it gains 2 hours in a month, near enough. One month after the autumn equinox, the sidereal clock will be 2 hours fast, two months after this equinox it will be 4 hours fast, and so on. Therefore, by counting the number of months since September 22 (the actual date can vary by a day or two in any year) and multiplying this number by 2 we find the approximate amount by which the sidereal clock is fast in hours. For each day after the 22nd of any month, we must add 4 minutes.

This makes a rather tedious and only approximate calculation. For one thing, the time of the equinox changes each year because there is not an exact number of days in the year. The odd quarter of a day is neglected for three years in our calendars, and we add an extra day to February every 4 years in leap year. In addition, our sidereal clock is gaining all the time, not in sudden 3·9 minute jumps at the end of the day. However for many purposes we can ignore these errors.

We could make another simplification without introducing too great an error. Half a month before the 22nd is about the 6th, when the sidereal clock will have gained one hour in addition to the time it has gained since September 22. For each week after the 22nd or the 6th of a month we could add 30 minutes to the difference. How will it work in practice?

What is the sidereal time at 10h 00m on October 29? Add 2 hours for the one month since September 22 and 30 minutes for the week since October 22. The sidereal clock will be 2h 30m fast, so the sidereal time will be about 12h 30m.

What is the sidereal time at 16h 47m on May 15? 7 months takes us to April 22, which makes 14 hours gained, to which we add 1 hour, because May 6 has passed. That gives a total of 15 hours so far. The 13th is one week later, so add another 30 minutes, and 2 days later we must add another 8 minutes. The sidereal clock is 15h 38m fast on 16h 47m. Adding these times we get 32h 25m from which we subtract 24 hours to get the sidereal time, 8h 25m. This means that a star of RA 8h 25m will be due south at 16h 47m meantime on May 15. Now we could make a correction for our longitude east or west of the central longitude of our particular time zone. to find when the star would be due south where we are.

Clocks in the sky

There are more convenient ways of finding the sidereal time, which we will now examine, but it is important to be able to make a rough estimate as described above. You can impress a friend one night by estimating the time by looking at the stars, but you will have to be able to estimate the difference between mean and sidereal time for that date. To do this you must learn to recognise the constellation of Cassiopeia. We will study the constellations, or groups of stars, in the next chapter, but we will start with this one because it so happens that one of the stars in Cassiopeia has an RA 00h, near enough, so it is almost on the same hour circle as the First Point of Aries.

Cassiopeia is easily spotted since the brightest stars in the constellation form a 'W', when beneath the pole (to the north), or an 'M' when above the pole. These stars are only 30° from the Pole Star so that, in theory, they are visible through the year from northern temperate latitudes. In practice, the constellation can occasionally be indistinct when it is immediately beneath the pole, that is when due north. Polaris itself is not a very bright star compared with some we can see, but there are no bright stars in its vicinity in the sky so that you should never have any doubt about which star it is. You should now be very familiar with your locality and know the directions which are north and south to within a few degrees without reference to your sundial or your other Sun shadow experiments. Polaris

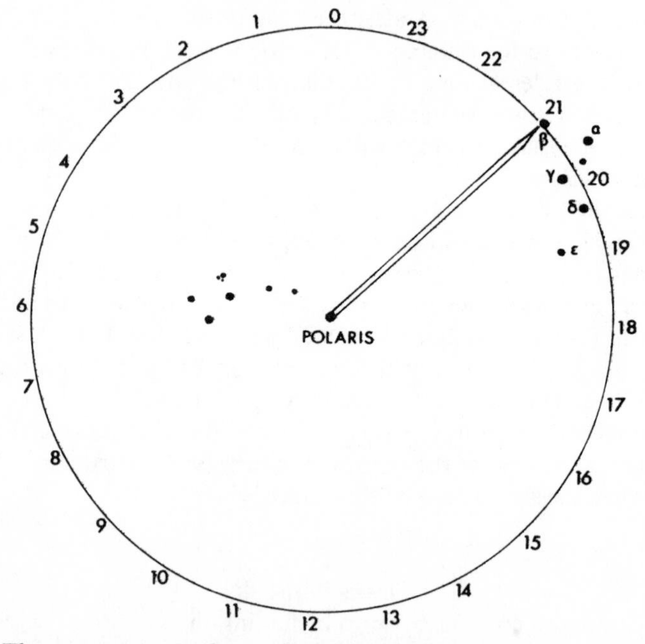

Fig 23 The imaginary 24 hour clock around the pole star with β Cas at the end of the hour hand which moves anticlockwise.

is due north, and its altitude is the same as your latitude. If you are uncertain, refer to Chapter 4.

Cassiopeia is on the opposite side of Polaris to The Plough, and about the same distance from it. When Cassiopeia is above the pole, you will find it high in the sky from moderate northern latitudes. The two stars which make the western arm of the M are clearly brighter than the others. The star at the end of the western arm, called β Cassiopeia has an RA of about 00h. When this star is due south then the sidereal time is therefore 00h (the RA on the meridian). This star describes a circular star trail around the pole, like all the others, in an anticlockwise direction, as shown in Fig 23. So if we imagine a large clockface centred on Polaris marked in 24 hourly divisions anticlockwise, β Cas. is the end of a sidereal hour hand, and it enables us to estimate sidereal time at a glance.

On February 5 β Cas. is seen to the left of Polaris. What is the time? If the star is exactly to the left, that is, due west of the pole, the star crossed the meridian 6 hours ago, so the sidereal time given by the imaginary 24 hour sidereal clock (Fig 23) is 6h. Now on February 6 the sidereal clock will be 9 hours fast, so on the 5th it will be 8h 56m fast. Subtracting 8h 56m from 6h (add 24 hours first) gives 30− 8h 56m = 21h 04m meantime.

Tables of sidereal time are given in a number of reference books. (See the suggested list in the Appendix.) A slightly more approximate table is given in the appendix to this book, but it will be accurate enough for any calculations you may want to make without a powerful telescope.

The Nocturnal

There is another 'clock in the sky' which uses the 'Pointers' in the Plough to give the mean time with a simple calculation, and you can work out how to use it in one of the exercises at the end of this chapter. Using these same stars, or indeed any star of known RA near the celestial pole, you can make a useful little gadget called a star nocturnal to tell the mean and sidereal time with surprising accuracy, and it can be used as a sidereal time calculator (a sort of circular slide rule) for any time in the year.

To make the nocturnal (see Fig 24), obtain a piece of finished wood, about 1cm × 2cm and 20cm long (the dimensions are not critical). Draw a thin straight line down the centre of one 2cm side. Now draw a circle of about 15cm diameter on a piece of thin card, and another circle on the same centre with a radius about 6mm less than the outer circle. Divide the outer circle into 24 equal parts by measuring 15° intervals from one radius, and mark lines radially at each position between the two circles. Subdivide each of these hours into quarters by striking off

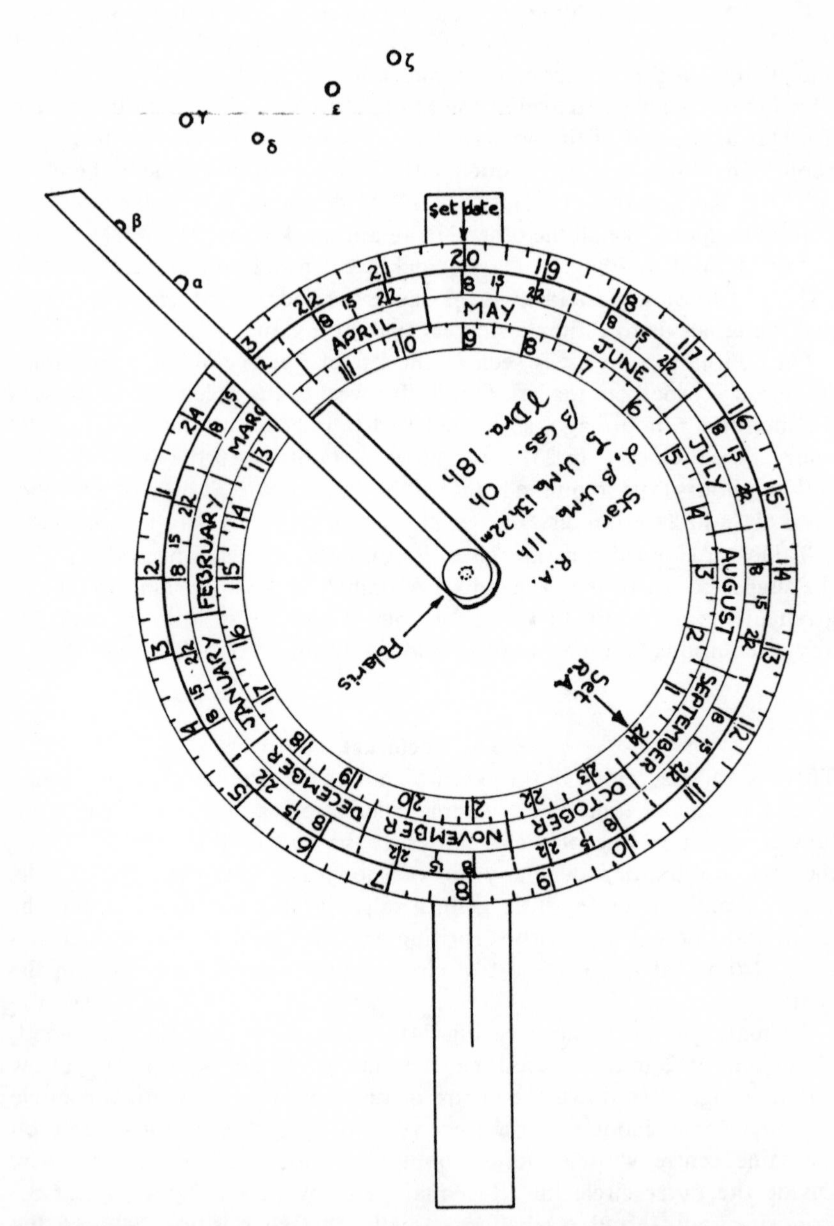

Fig 24 The star nocturnal, set to read the time by the Pointers (α and β Ursa Majoris) on May 9. UT is read on the outer scale, and is 23h 00m.

faint arcs on the outer circle with your compass. Number the hour lines from 1 to 24 (or 00h) anticlockwise. Now you can cut out the disc around the outer circle.

Draw another circle on a new piece of card, with a diameter which will just fit inside the inner circle on the first disc. Draw three more circles concentric with the first, and each of radii something under 6mm less than the one outside. This will form three narrow circular scales inside each other. On the innermost scale, mark out another 24 hour scale at 15° intervals, together with the quarter hours and number anti-clockwise, just like the scale on the larger disc. On the same radius as the 24h mark on the inner scale, mark 22 on the outer scale (this will be September 22). Opposite 1h on the inner scale, mark 8 on the outer scale. Opposite the 2h mark, write 22 again, and number the outer scale in this way right around the dial with '22' opposite each even hour, and '8' opposite each odd hour. Now write '1' and '15' at the half hours, so that scale reads 22, 1, 8, 15, 22, 1, 8 and so on clockwise. At each '1' on the outer scale draw a radial line across the middle scale, giving twelve divisions, one for each month which you can now write on the middle scale starting with September between 1h 30m and 23h 30m, followed by October, also going clockwise. The finished dial can now be cut out.

Cut a strip of the card about 2cm wide and 15cm long and draw the centre line down its length. Make a mark on the centre line about 12mm from one end, and push a drawing pin through the line at this mark. Push the pin through the centre of the smaller diameter disc and make a mark on the strip just inside the inner scale of the card. Carefully cut an absolutely straight line down the centre line of the strip up to the mark you have just made, and cut half way across the strip to the centre line at this point. Now push the pin through the centre of the larger disc and into the centre line of the piece of wood so that the end of this line is just visible above the larger disc. To finish off the nocturnal, swing the strip down until it is on the lower part of the wooden handle and trim off any excess from the arm.

To find sidereal time with this instrument, turn the date on the inner dial to 24 hours (or 00h) on the outer dial. Using the arm as a guide, the time on the inner scale of the smaller disc in UT can be read against the sidereal time on the outer disc.

To use the nocturnal to tell the time, the centre of the discs must be held in front of Polaris; although you can't see the star itself, it is surprising how accurately you can judge its position with the Pointers as guides. You may prefer to use a thick card strip for the handle and secure the discs and arm to this with a small circular rivet so that you can actually see through the centre to line up on Polaris.

To find mean time from a star near the pole, set 24h or 00h on the inner dial against the RA of the star on the outer dial. The dials must be kept in this relative position for all measurements using that particular star, so, turning them both together, set the present date against the top index, that is, the centre line of the handle above the discs. Now hold the nocturnal with the handle vertical and with Polaris behind the centre of the discs, and turn the arm until the edge on the centre line is just touching the star. Now read the mean time on the outer scale where the same edge of the arm crosses it.

You will notice that when the two discs are set up for a given star, the date opposite 24h on the outer dial is the date when that star transits at midnight, 00h. In the case of the Pointers, both stars have RA 11h 00m, and you have two stars to line up on the edge of the arm, as is shown in Fig 24. The RA of some other circumpolar stars which you could use, are as follows: ζ UMa (Mizar), 13h 22m; β Cas 00h; γ Dra 18h. We will return to the numbering of the stars in the next chapter.

A thorough understanding of star position measurement and sidereal time is of the utmost importance to your grasp of the elements of positional astronomy and its practical applications, or for any observational work which is to be more than mere sightseeing around the sky. So work through the exercises which follow very carefully. On January 21, the RA of the Sun at noon was 20h 11m. What was the value of the equation of time?

Using the sidereal time tables given in the appendix, we find that the sidereal clocks are 7h 57·9m fast on UT at 00h on this date. By 12h 0m, the sidereal clocks will have gained another half day's quota, 2 minutes, so the total difference will be 7h 59·9m. The sidereal time at noon is, therefore, 19h 59·9m. The Sun will not reach the meridian for 20h 11m − 19h 59·9m = 11·1 minutes, so the value of the equation of time is −11·1 min, the minus sign indicating that the Sun is slow.

EXERCISES ON CHAPTER 3

1 The RA of the Pointers is 11h. An imaginary clock face around the pole is marked 0 to 23 anticlockwise, with 00h above the pole. When the 'hand', ie the Pointers is at 16h on this dial on November 5, what is the mean time? From this exercise you can work out the rule for using the Pointers as an alternative 'clock in the sky'.
2 Venus sets due west exactly on May 19. What is its RA and Dec, roughly?
3 Saturn is at RA 5h 06m. On what date will it be due south at midnight?
4 What is the RA of the Sun at noon of 1976 November 13?
5 The constellation of Scorpius has a Dec of about −15° and RA of 16h (very roughly). From this data can you estimate a good time of the year to look out for this constellation from northern latitudes at about midnight?
6 New Moon occurs on April 13. What is the approximate RA of the Moon 7 days later?

4
Stars and Constellations

In the last chapter we learned how the sky is measured in terms of a large globe with celestial co-ordinates of right ascension and declination. This is an essential background to learning your way about the sky, and it should soon become second nature always to have an idea of the sidereal time and, with that, an idea of the positions of the most important stars and groups of stars in the sky.

For those readers who are complete newcomers to the subject it must be explained that stars are recognised usually by their position in a group. The groups, called constellations, are fairly obvious in many cases—such as the famous group of seven stars known as the Plough. To find the Plough—which is not a constellation but merely the prominent stars in the constellation of The Great Bear—estimate when the sidereal time is 11h for the current date, and the leading edge of the Plough will be south of the pole and almost overhead at latitudes of 40 to 60° N. If 11h ST happens to fall in the middle of the day, add 12h and look for it due north at that time. The Plough extends over an angle of some 25° and its shape is shown in Fig 24. This also shows how the two leading stars have the same RA, near enough, so that they point towards the north celestial poles. Being within 30° of the north pole, these 'Pointers' are a sure indicator of Polaris, the Pole Star.

If we follow the line drawn through the pole by the Pointers to about the same angle on the other side of the pole, we come to the easily recognised capital W or M, the heart of the constellation Cassiopeia.

The stellar constellations are usually referred to by their Latin names, thus the Great Bear of which the stars of the Plough are the most easily recognised is Ursa Major. Polaris is at the end of the tail of the Little Bear, Ursa Minor, shown in Fig 25. The names of the constellations are as old as recorded astronomy in many cases, and most of them represent a very imaginative interpretation of the star patterns. It is not at all difficult to see how a few of them came to be named, such as the Scorpion, the Swan, the Dolphin, the Lion, and the Arrow; but the Charioteer, the Water Carrier, the Greater and Lesser Dogs, and many others demand a stretching of the imagination!

It is pointless trying to learn the pattern of all the stars within the constellation boundaries; the best way to start is to become familiar with the most prominent stars of the constellations visible at different times of the year. Once you have mastered the main constellations you will be able to learn the fainter, less obvious groupings easily.

Within each group of stars, the individual stars are numbered by using the Greek alphabet. The stars are usually lettered according to their relative brightness, starting with α for the brightest, β for the next, and so on. The dimmer stars are often lettered or numbered using the conventional alphabet and numerals. The complete Greek alphabet is given in the table below, and it is worthwhile learning at least the first ten or so, since these letters are used for the brighter stars with which we shall be concerned for the most part.

α	alpha	η	eta	ν	nu	τ	tau
β	beta	θ	theta	ξ	xi	υ	upsilon
γ	gamma	ι	iota	o	omicron	ϕ	phi
δ	delta	κ	kappa	π	pi	χ	chi
ε	epsilon	λ	lambda	ρ	rho	ψ	psi
ζ	zeta	μ	mu	σ	sigma	ω	omega

The illustrations in this chapter show the brightest stars of the constellations visible in the northern sky, and down to declination $-40°$. From our earlier work we know that we can never see stars further south than declination $(90-\text{lat})°$ south and, in practice, the stars which manage only to peep over the southern horizon will not be visible either. We have only to think of the dull red Sun on the horizon to realise that even the brightest stars will be dimmed beyond visibility when they are very low in the sky.

Except for Fig 25, which shows the stars closest to the north celestial pole, the stars are described as if the central hour circle of RA were on the meridian, so that the lefthand side of these charts is to the east, and the righthand side of the chart is to the west.

The brightness of a star is expressed as a *magnitude*. The smaller the magnitude number, the brighter the star is; thus a magnitude 2 star is brighter than a magnitude 3 star, but dimmer than a magnitude 1 star. Star magnitudes are based on the ratio of their brightnesses such that a *difference* in magnitude of 5 represents a ratio of brightness of 100 times. Thus a magnitude 1 star is 100 times brighter than a magnitude 6 star. It works out that for a difference in magnitude of 1, the ratio of brightness is 2·512, so we can say, with sufficient accuracy for our purposes, that a magnitude 1 star is $2\frac{1}{2}$ times brighter than a magnitude 2 star.

Stars are normally referred to as being of first, second, third magnitudes, and so on. In general, 'first magnitude' stars are those brighter than magnitude 1·5, 'second magnitude' stars are between 1·5 and 2·5, and so on. The very brightest stars have magnitudes of about 0, and two are so bright that they must be given magnitudes of less than 0: these are shown with a negative number. The brightness of a planet is also expressed as a magnitude, and frequently these are also negative. Venus, for example, reaches magnitudes brighter than −4, which means it is 100 times brighter than a star of magnitude 1. Some stars vary in brightness, and so are given a range of magnitudes. Without a telescope, and in the very best possible sky conditions, an observer with good eyesight may just be able to see a sixth magnitude star.

Fig 25 The circumpolar stars at latitude 40° north, shown at the spring equinox. The scale from 6h to the pole shows the diameters of the Dec circles of 90°, (the pole), 80°, 70°, 60°, and the outer circumference 50°.

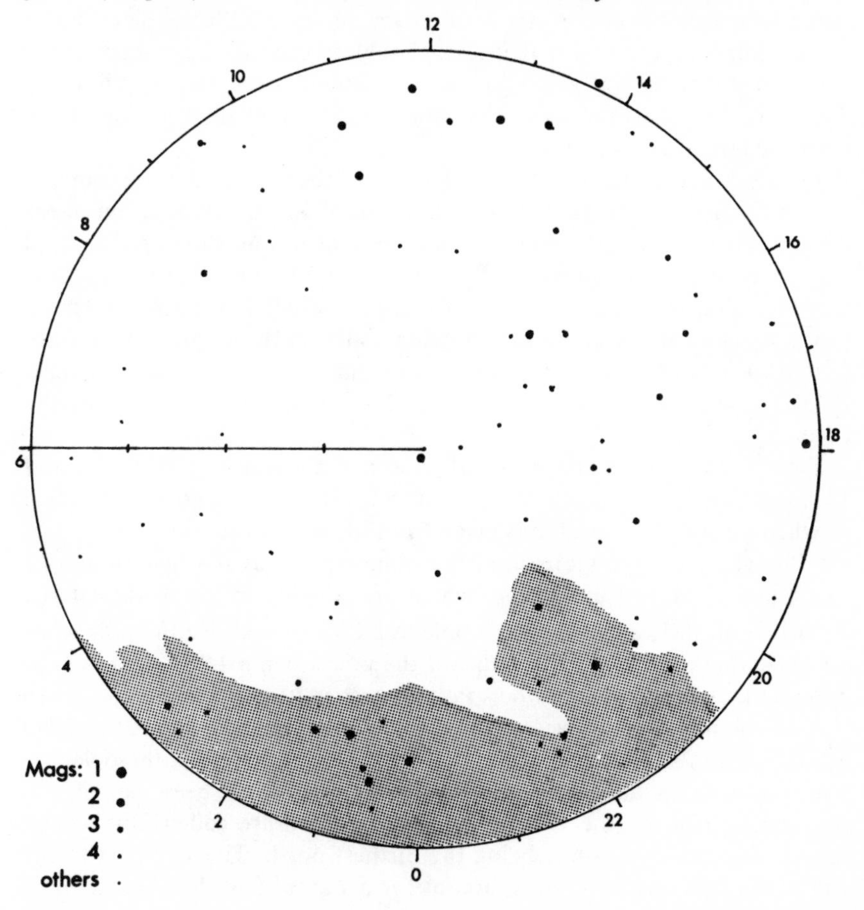

We have seen that the brightest, and in some cases some of the interesting stars which are not so bright, have been given names. These date back to the very earliest days of astronomy and in most cases were given to the stars by the ancient Greeks and Romans, and the Arab astronomers of the middle ages.

The names, and other information such as positions, are given in the text, rather than on the charts themselves. This sort of information on the star diagrams only serves to confuse the star patterns. Neither have the stars been joined by lines. The most important thing to do is to familiarise yourself with the star patterns as *you* see them, not with somebody else's ideas of what they look like.

An overlap between each chart enables you to relate the stars which happen to fall on the boundaries of the charts to those either to the east or west. The different magnitudes of the stars, which are shown down to magnitude 4 on the charts, are depicted by circles of different sizes, but it is most important to realise that even though to the naked eye some of the brightest stars may appear bigger than the dimmer ones, this is an illusion. The stars are all so far away that none shows more than a pinpoint in even the largest telescopes on Earth.

Fig 25 shows the region of the pole, and all the stars in this diagram are visible throughout the year at latitude north of 50°, the declination of the outer circle enclosing the diagram. The scale at RA 6h shows the radii of the circles of declination within the diagram at 10° intervals to the pole at Dec 90°. You will notice the end of this line (which is the sole exception to the remarks above about not drawing scales on the diagrams) does not quite touch Polaris, since this star is not quite at the true celestial pole. Polaris is at about RA 2h 05m and Dec 89° 05′ nearly 1° from the pole.

Polaris is therefore on the meridian above the pole at ST 2h 05′, and 12 hours later it is beneath the pole at ST 14h 05m. A good indication of when it is on the meridian is given from the star at the end of the tail of the Plough, η Ursa Majoris, and the dimmest star at the eastern end of Cassiopeia's 'M', ε Cassiopeiae, which are roughly in line with Polaris. Polaris is on the same side of the pole as ε Cas, so is above the pole when ε Cas is above the pole, and beneath the pole when η UMa is above the pole. Thus you can judge when Polaris best indicates true north—when it is on the meridian above or below the pole, and when it is at the exact altitude to measure your latitude—when it is east or west of the pole.

In Ursa Major, all the 'seven stars' (the Plough) have been named, but the best known are the Pointers, α and β, which are called Dubhe and Merak respectively, Dubhe being the furthest north. The star next to the end of the tail is Mizar, and is famous for being the first double star to be

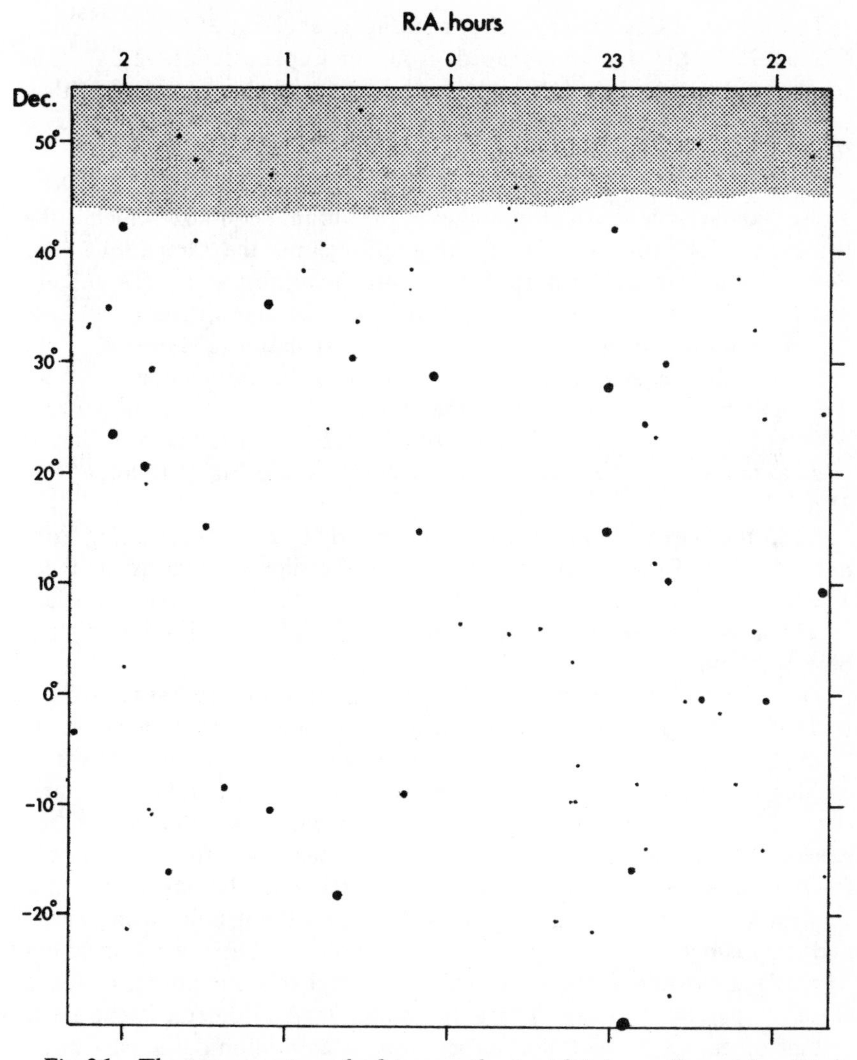

Fig 26 The autumn stars, looking south at midnight on September 22.

recognised as such. Through a telescope, we see that Mizar is two stars, but their angular separation is only 14·5″. Very close in the sky to Mizar is a much fainter companion, Alcor, which can be seen with the naked eye quite easily, and is shown in the diagram.

From Polaris itself we can see the rather inconspicuous constellation of Ursa Minor, the Lesser Bear, curving away from the pole towards the tail of its bigger brother. The brightest stars in Ursa Minor (UMi) form a little rectangle with a tail, at the end of which is Polaris.

To the east of Cassiopeia as shown in Fig 25 at about 21 to 23h RA and Dec 60–75° is the diamond-shaped group in Cepheus. Eastwards again to about RA 18h, Dec 55°, is a smaller diamond-shaped group with a tortuous trail of rather faint stars which winds its way between the two Bears. This is the Dragon, Draco.

Fig 26 shows the stars to the south of Cassiopeia, centered on 00h RA. These stars are due south at midnight at the autumn equinox, which is the best time to look for them. Dominating this region is the 'Great Square' of Pegasus. The two eastern stars in the square are almost at 00h RA and are useful for instant estimates of sidereal time. The star in the north east corner of the square is in fact not in the constellation of Pegasus at all, but is *α* Andromedae (we use the genitive case of the Latin names when we use the constellation name with the star numbers, but it is not worth while trying to learn these all at once just yet). Pegaus itself is the group of stars to the west of the Square, and extends well into Fig 31 to about RA 21h.

Andromeda extends a long arm of three fairly bright stars ,starting from *α*, to the east of the Great Square, and curving slightly towards the pole. The third star, at RA 2h Dec 42°, is a famous double star, *γ* Andromedae, in which the two component stars are of incredibly beautiful contrasting blue and gold.

The centre star in the bright Andromeda trio is *β*. Two rather faint stars, both of fourth magnitude, are shown on the diagram curving northwards from *β*. Close to the nothernmost, *υ*, is one of the best known objects in astronomy, the Great Nebula in Andromeda. This is a galaxy of stars, relatively close to our own galaxy compared with the millions of other galaxies which have been found. It can be seen close to *υ* and, just a fraction to the west, as a very faint hazy patch with the naked eye on a very dark ,clear night. With a pair of binoculars the nebula is very clear, and surprisingly large. Only a very few nebulae (a Latin word meaning 'clouds') are visible to the naked eye. All except this one are in, or fairly close to, our own galaxy. The great galaxy in Andromeda is the most distant object visible to the naked eye, about 2¼ million light years away from Earth.

Almost in a line below the western side of the Great Square is a very bright star which, at Dec −30°, really belongs to the southern hemisphere observers, but close to the autumn equinox nothern observers are allowed a glimpse. This is *α* Piscis Austrinus (the Southern Fish) which has the poetic name Fomalhaut. It has a magnitude of 1·29, and so is not a leader in the brightness league, but is well worth looking for.

To the east of the Great Square, and below *γ* And, is a fairly bright star called Hamal with a fainter companion to the south west. Hamal is the star α in the constellation of Aries. This is the constellation through which the ecliptic passed and crossed the equator at the spring equinox, the First Point of Aries, so many centuries ago. The band of sky containing the constellations like Aries through which the ecliptic passes is known as the Zodiac, which means circle of animals, because most of the twelve zodiacal constellations have animals' names, such as Aries, the Ram.

RA 00h 00m Dec 0° is now in the constellation of Pisces, the Fishes, the stars of which are to be found south of the Great Square but which are all about fourth magnitude or less. Between Hamal and γ And, is a faint, but easily spotted constellation of three stars in a triangle called, with surprising originality, Triangulum!

Winter splendour

As autumn gives way to winter, the sky assumes an aspect unsurpassed at any other season. Late in the November evenings, one of the most spectacular constellations in the heavens shoulders his way over the horizon almost exactly in the east. This is Orion the Hunter, a large bright constellation of an unmistakable shape, clearly suggesting a striding figure complete with a belt about his waist from which a sword can be seen suspended. The sword is not shown completely in Fig 27, because it is composed of very faint stars—but they must be among the easiest faint stars to see in the sky. Orion straddles the equator (which is why it rises due east) at RA 5–6h. It contains two very important first magnitude stars. The first, α Orionis, is called Betelgeuse. Its magnitude varies between 0·5 and 1·4 irregularly. Since you can easily detect this change in brightness with the naked eye, it is always worth locking at Betelgeuse to see if it has changed since you last saw it.

Betelgeuse is Orion's left shoulder, and β Orionis, the second very bright star, is his right foot. This star, called Rigel, has a magnitude of 0·3, and is a brilliant white. A faint arc of stars to the west represents a skin hanging over the hunter's arm, stretched out towards the Bull. At the end of his sword is a faint star, θ, and immediately above is a misty patch, clearly visible on a clear night with the naked eye and very clear in binoculars. The misty patch is another nebula, which rivals the nebula in Andromeda for fame, and certainly exceeds it in beauty. The Great Nebula in Orion is a cloud of glowing gas many light years across in which, it is generally believed, new stars are being formed.

Immediately beneath Orion's feet is a small constellation, Lepus, the Hare. Following the line of Orion's belt towards the north west we find another first magnitude star at one end of a V of faint stars. The V is the horns of Taurus, the Bull, one of the zodiacal animals. The bright star, α

Tauri, is called Aldebaran. The V extends well out to the east over Orion, terminating in a bright star at about RA 5h 25m, Dec 28°, which is β Tauri, and a fainter star at about RA 5h 35m, Dec 21°, ζ Tauri.

The smaller, more obvious V in which Aldebaran is found contains very many more stars than you can see with the unaided eye, but even binoculars reveal much more of this famous cluster of stars, which is called the Hyades. Aldebaran itself is another reddish star, magnitude 1, and so forms a useful

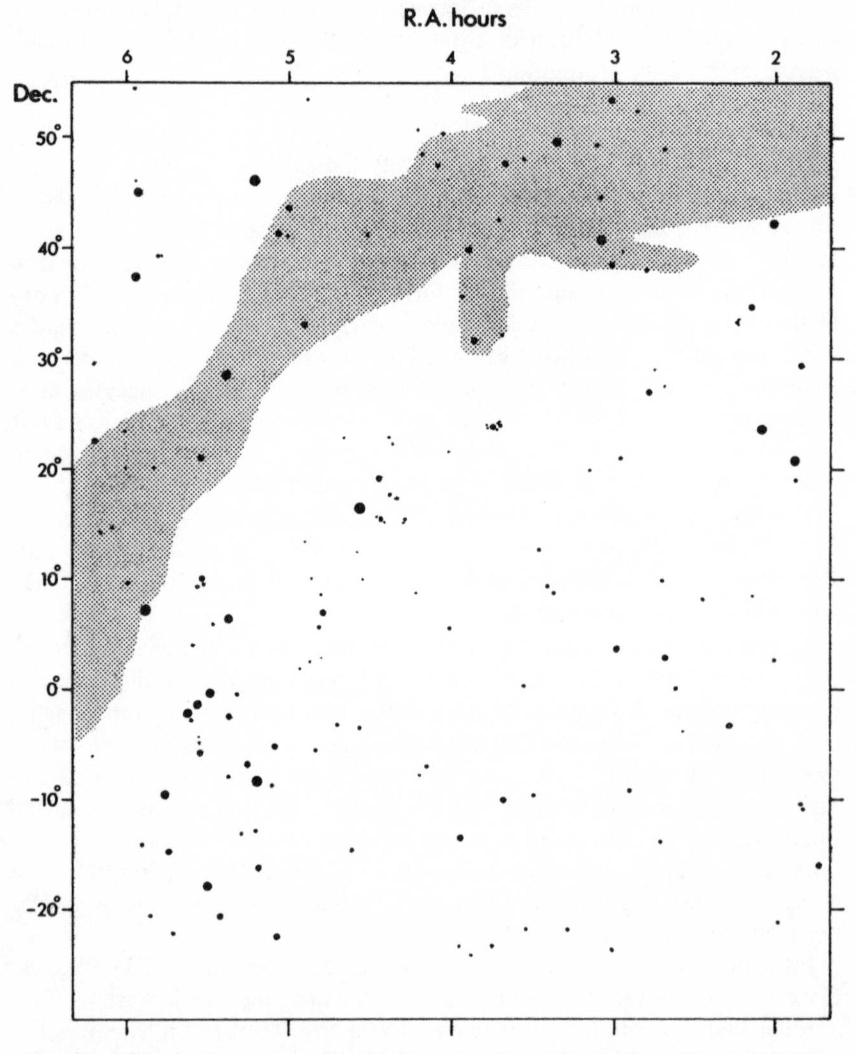

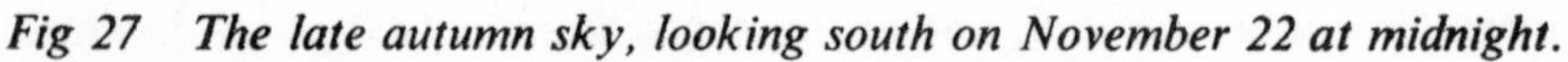
Fig 27 The late autumn sky, looking south on November 22 at midnight.

comparison for estimating the magnitude of the variable Betelgeuse.

Continuing the line from the belt of Orion through the Hyades we come to a well known star cluster, the Pleiades. These are to be found at RA 13h 40m and Dec 24°. Only one of the Pleiades is brighter than fourth magnitude, and keen eyesight is needed to see seven or eight without optical aid. With a pair of binoculars, the region is found to be packed with stars, and is an unforgettable sight. Being high in the sky in northern latitudes, the cluster is visible from quite early in the autumn until well into the following spring.

Another large and easily identifiable constellation is to be seen throughout the winter. Sitting on the far end of the northern horn of the Bull, β Tauri, is Auriga, the Waggoner or Charioteer. The constellation in fact bears no resemblance to either, and is more easily seen together with β Tauri as an unequal sided pentagon. At its northern apex is yet another of the sky's brightest stars, α Aurigae, called Capella. This has a magnitude of 0·2, and is quite yellow compared with, say, Rigel. To the south of Capella is a triangle of fainter stars called the Kids.

Immediately preceeding Auriga in his journeys through the heavens, that is, to the west, and roughly on that same line of Orion's belt through Aldebaran and the Pleiades, is the constellation of Perseus. He is alongside Andromeda with whom he is linked in mythology, and he carries the head of Medusa, the Gorgon. Perseus is most easily seen as an arc of four stars; second magnitude α, with third magnitudes γ and δ to the north west and south east, and fourth magnitude η at the northern end of the arc. This star η is too far north to appear on the chart in Fig 27, but is to be found in the circumpolar map, Fig 25, at RA 2h 47m, Dec 56°. This star is another interesting double of contrasting colours.

Below α, and almost forming a right angled triangle with γ Andromedae (which appears on the western edge of the chart) is a star at RA 3h 5m, Dec 41°. Normally, this star is not quite as bright as α, and clearly brighter than its fairly close companion to the south, ρ. This is Algol, β Persei, the eye of Medusa, also called the Demon Star, because its strange behaviour suggested to the ancient astronomers a most sinister object. Regularly every 2½ days, it fades from its usual magnitude 2·3 to 3·7, taking only 5 hours to do so. At its minimum, it is only just a little brighter than its companion ρ. After 20 minutes it steadily regains its former brightness over the next 5 hours. This extraordinary behaviour is due to the fact that Algol is a binary star, made up of two stars very close together which rotate about their common centre of gravity. One star is very much dimmer than the other, and it so happens that its plane of rotation points towards us, so that as the dimmer star circles about the bright companion, it partially obscures its light from us. The whole cycle of events takes 2 days 20 hours and 49 minutes. There is no mistaking Algol at minimum and it is an event to watch for. You can compare it with *α* Persei, or with nearby *ρ*, but the star *ρ* is also variable, changing from magnitude 3·3 to 4·1 irregularly, so be careful to decide which of the two has changed in brightness.

This region contains another famous variable star, which is even more remarkable than Algol, but not quite so easy to see. This is *o* Ceti, called Mira, 'The Wonderful'. This just appears on Fig 26, but can be seen more clearly in relation to the surrounding stars in Fig 27 at RA 2h 17m Dec −3° 12′. This star can be found well to the south of Hamal and its companion in Aries. It is also in the faint line of stars stretching westwards from the base of the V in Taurus. It is shown on the chart as about third magnitude, but with a 'hole' in the centre. This is because at maximum it can be as bright as magnitude 1·7, while at minimum as low as 9·6, although its range is usually less than this—typically from magnitudes 3 to 8. This irregular variation takes about 330 days to complete a cycle. At minimum it is quite invisible except in a telescope of reasonable size. Then it becomes visible to the naked eye for about six months before once more sinking out of sight. This fantastic increase of brightness, some 1500 times, is thought to be due to immense outbursts of gas from the star.

The Milky Way, which is formed of the millions of stars in our Galaxy of stars, passes through the middle of this region of bright winter constellations to add the final touch. The Galaxy in which our Sun is a rather insignificant member, is a huge disc of stars some 100,000 light years across and containing about as many millions of stars. The central plane of this disc is called the galactic equator, and this encircles the celestial sphere along the middle of The Milky Way. Fig 32, towards the end of this chapter, indicates the paths of the galactic and celestial equators, together with the ecliptic in more detail.

The region of the sky shown in Fig 28 is dominated by a single star, Sirius. This is the brightest star in the sky, north or south, with a magnitude of −1·58, and RA 6h 44m, Dec −16° 40′. Its nearest rival is a southern constellation star, Canopus, which strangely enough would also appear on Fig 28 if it were extended southwards to Dec −53°, where it would be found almost due south of Sirius. Canopus has a magnitude of −0·86. Sirius is the star *α* in the constellation Canis Major, the Larger Dog. The rest of the constellation includes four other bright stars, all close.

The Larger Dog trails faithfully at the heel of his master, The Hunter, and the Lesser Dog, Canis Minor, follows much higher in the sky. Canis Minor is easily found; a pair of stars with no others easily seen in the immediate vicinity, to the east of Betelgeuse. The star *α* Canis Minoris is among the ten brightest stars in the sky. Called Procyon, it is found at RA 7h 37m, Dec 5°.

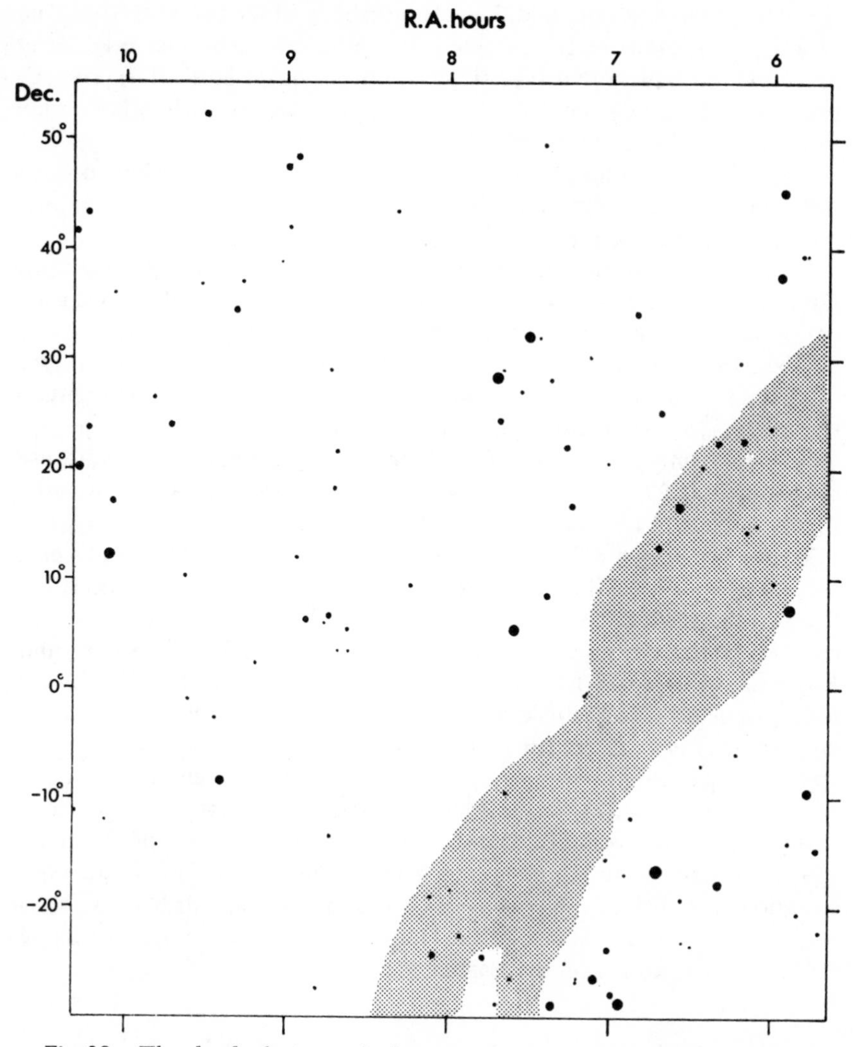

Fig 28 The dead of winter, looking south at midnight on January 22.

To the north of the two prominent stars of Canis Minor is another pair of stars resembling the Lesser Dog in separation and orientation, but comprised of two first magnitude stars. This is the zodiacal constellation Gemini, the Twins. The ecliptic passes to the south of these two bright stars at some two thirds of the distance from Procyon. At first glance the Twins appear aptly named, and yet there is little similarity between them.

Castor, α Geminorum, and the most northerly of the two at RA 7h 31m, Dec 32°, is magnitude 1·58. Pollux, β Geminorum, is in fact brighter at magnitude 1·21 (note that the brighter of the Twins is labelled β). Pollux's position is RA 7h 42m, Dec 28°. Castor is also distinctly whiter than Pollux, which has a yellowish colour.

Two fairly faint trails of stars lead away from Castor and Pollux towards the south west and Orion, and if this line is followed in the opposite direction we find the Plough.

On the equator to the east of Canis Minor in Fig 28 is an inconspicuous group of stars which are at the head of the constellation Hydra. This is an enormous constellation which starts at the little group at about RA 8h 30m, Dec +5°, wends its way south to α Hydrae at RA 9h 20m, Dec −8°, then on to the south west with a winding trail of fourth magnitude stars down to a declination of some 30° south where it crosses the sky to RA 15h (only about two of the stars in the remainder of Hydra are shown in Figs 29 and 30 across which it meanders, too far south to be of any interest to northern observers).

To the north of the Hydra's head is an apparently rather empty part of the sky; but here, if you look hard, is the constellation of Cancer, the Crab. Cancer is another of our zodiacal animals, and it is to be found roughly in the centre of the triangle formed by Pollux, Procyon, and Regulus. Another aid to finding it is to look to the west of the Lion's head (see below). To the naked eye, Cancer is unremarkable, and contains no stars greater than fourth magnitude. Two of these fourth magnitude stars, ν and δ, are to be found in the centre of the constellation, roughly half way along the line from Pollux to Regulus. Almost between these two faint stars is a beautiful group of stars called Praesepe (the Beehive), which are well worth studying in a pair of binoculars. This is the constellation in which the Sun was once found furthest north (this is no longer true due to precession of the equinoxes) and hence it gave its name to the northern tropic on the Earth.

Spring stars

In the region of RA 10h, Dec +20° in Fig 28 is the head of the Lion, sometimes called the Sickle, and which is very easily recognised by its appearance as a backward question mark. At the bottom of the question mark is a star of magnitude 1·34, Regulus, α Leonis. Regulus is a most useful celestial marker; its RA is 10h 6m and Dec +12° 13′, which puts it very close to the ecliptic. In fact, the ecliptic passes less than one degree south of Regulus. Being near to the ecliptic, close to which the Moon and planets move, Regulus is comparatively often involved in close and spec-

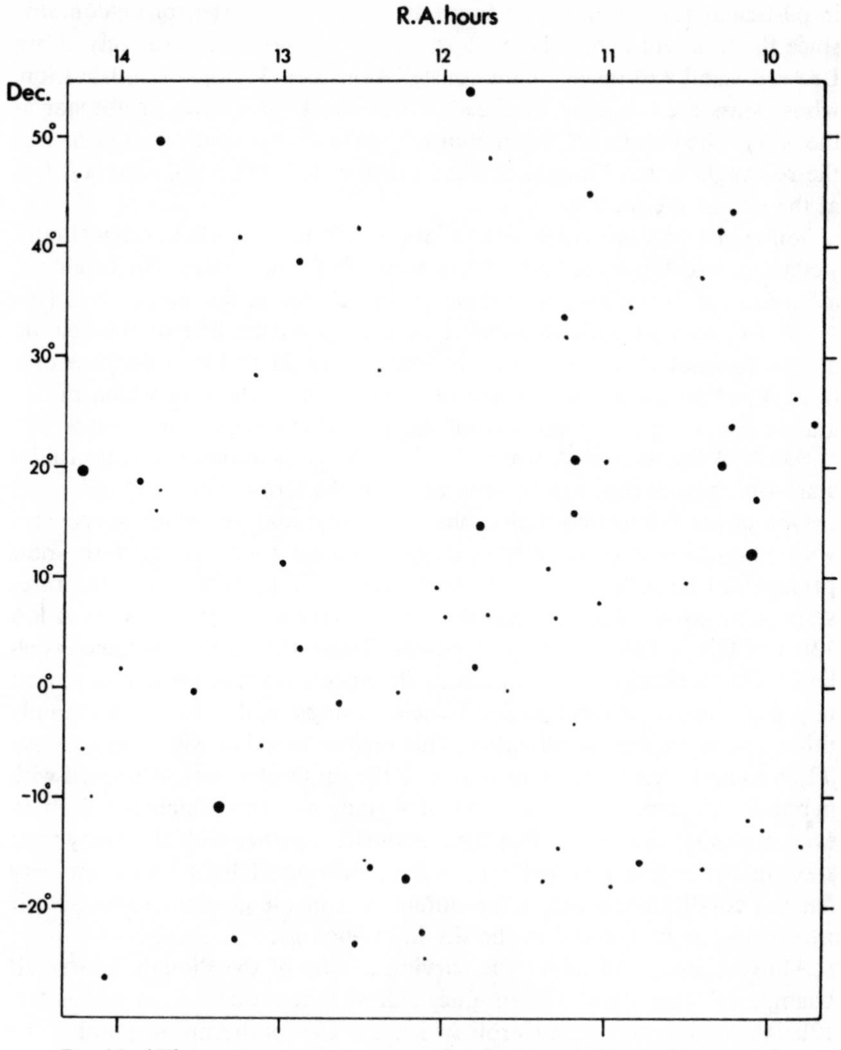

Fig 29 The spring equinox, at midnight March 22, looking south.

tacular conjunctions with the Moon and planets, and on more rare occasions, can be completely covered by one of them. Such an occasion is called an *occultation*.

The whole of the constellation of Leo, also a zodiacal animal, is shown in Fig 29. This constellation is easily seen as the Lion it represents. At its tail is the star Denebola (β Leonis) which at RA 11h 45m is close to the hour circle passing through the autumn equinox at 12h. When Leo, and in particular Leo's tail, is due south at midnight, it is the spring equinox, since the Sun will be on the equator at RA 00h 00m. Immediately above Leo is a small and insignificant constellation, Leo Minor, the Little Lion, whose stars are too faint to show on this chart. The fairly bright star at the top of the chart at RA 11h 50m is γ UMa. in the south west corner of the rectangle in the Plough, and the other at RA 13h 45m, Dec 50°, is η at the end of the handle.

South of Leo's tail is the constellation of Virgo, another zodiacal constellation, which is large and yet has only a few bright stars. The brightest, α Virginis, is Spica, also very close to the ecliptic at RA 13h 22·5m, Dec −10° 54′, so that with Regulus it clearly shows the line of the ecliptic across this part of the sky. Spica is magnitude 1·21 and is at the base of a large Y, of which γ is at the centre, η and β form the arm which points almost due west, and δ and ε form the arm which points northwards.

South of the westward arm of the Y in Virgo is a quadrilateral of faint stars which is all that can be seen easily of the Crow, Corvus.

One of the faintest constellations in the sky, and yet which is crowded with interesting objects telescopically, is to be found (only with some persistence) by following the northern arm of Virgo's Y. After the three stars in Virgo, the line is continued by two very much fainter stars at RA 13h and Dec +18° and 28° respectively. These are α and β of the constellation Coma Berenices, which means Berenice's Hair. A third star, γ, west of β and forming a right angled isoceles triangle with β and α is the only other star to be seen at all easily. This region, together with that between β Leonis and ε Virginis, immediately to the south of Coma, is packed with nebulae—an incredible collection of distant galaxies which are faint in even moderate telescopes. But these nebulae, together with the many faint stars in the region give the area a faint dusting of light under the very darkest conditions, which is the shimmering of the glorious locks of hair which were immortalised in the sky in mythology.

Above Coma, and under the curving handle of the Plough, is a small triangle of faint stars. The brightest, third magnitude α is at about RA 12h 54m, and is called Cor Caroli, which is at the southernmost point of the triangle. The constellation is Canes Venatici, the Hunting Dogs. To the extreme east in Fig 29 at RA 14h 13m, Dec +19° is another of the brightest stars in the sky, Arcturus, which can be seen on the western side of Fig 30.

Fig 30 shows the stars which are south at midnight at the end of May. Arcturus is in the tail of a kite formed of four widely spaced stars centred at RA 14h 50m, Dec +35°. This constellation is the Ploughman, pushing the Plough westwards around the pole. His name is Bootes and he is easily found in the spring and early summer from α Bootis, Arcturus, which is in the curve of the Plough's handle continued down the sky. Arcturus is

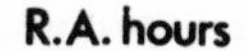

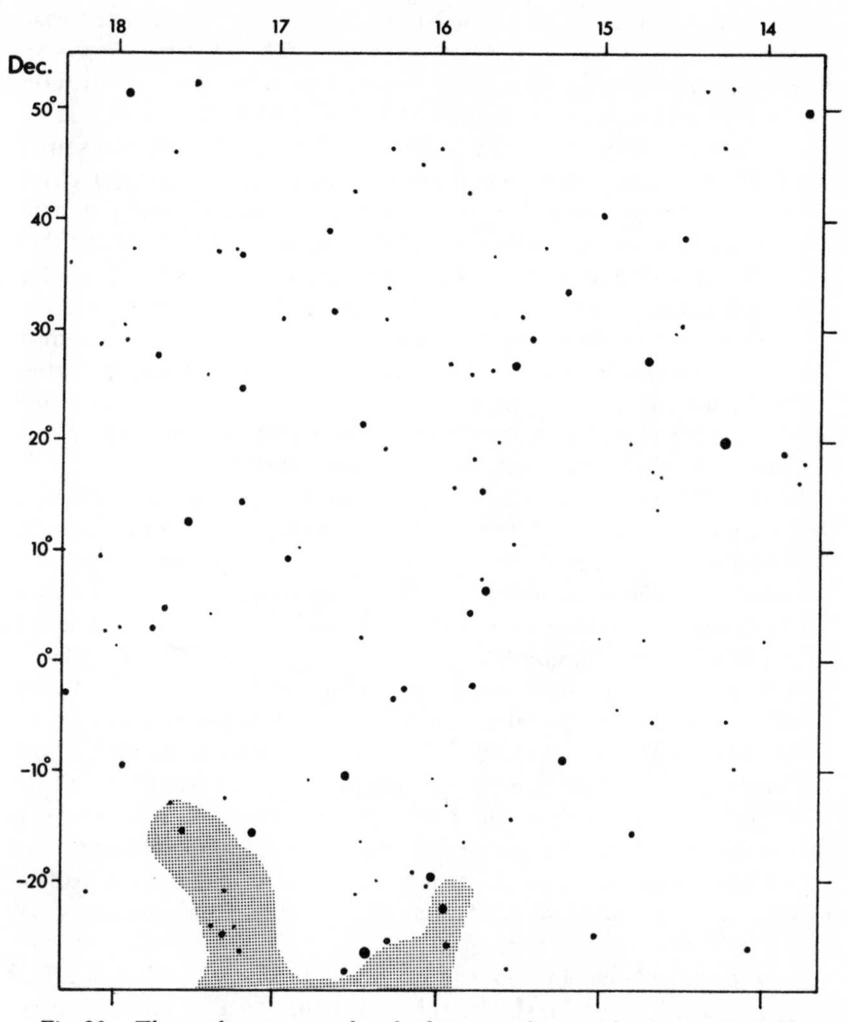

Fig 30 The early summer sky, looking south at midnight on May 22.

yellow in colour and of magnitude 0.2.

The star at the eastern corner of the kite, δ Bootis, is a double star of contrasting colours which may just be seen in a pair of very good binoculars. Just to the southeast of this star is a little semicircle of faint stars, the Northern Crown, Corona Borealis. This has one star which just about qualifies to be second magnitude, an adjacent third magnitude star, and the remainder are all fourth or lower magnitudes.

Eastwards again from Corona and not quite south of Draco's head is the fairly faint, but distinctive constellation of Hercules. This is most easily recognised by a trapezoidal group centred at RA 16h 50m, Dec +35°, extending curving arms from each corner. The north west arm is rather like a bent arrow, turning to point at Corona, while the north east line curves gently northwards to the head of the Dragon, Draco. The stars in this four sided figure, starting at the north west corner and going clockwise are η, ζ, ε and π, and on a very clear, dark night if you look at a spot on the line from η to ε, on the western side of the trapezoid, and about one third of the distance from η, you should see a small, faint, misty patch, another naked eye nebula, the Great Globular Cluster in Hercules. If you examine this cluster through a pair of binoculars you will see a tiny round misty patch. In a small telescpoe this cluster can be resolved into very faint stars. It has been estimated that there are over half a million stars packed into this, and other globular clusters. To help you to identify the exact spot on the chart, the cluster is at RA 16h 40m, Dec +36½°.

South of Hercules is another sprawling constellation which is ill defined and composed mostly of fainter stars. On the diagram it can be seen just below the southern limits of Hercules at about Dec 10°, and above the bright southern constellation Scorpius. This equatorial constellation is Ophiuchus, the Man with the snake. The snake has been rendred fairly harmless as it is split into two parts, the constellation of Serpens Caput, the Snake's Head, and Serpens Cauda, the Snake's Tail. You will have to become quite familiar with many of the faint stars in this region before you will be able to identify the Snake's anatomy. Ophiuchus is a huge incomplete ellipse, the easiest part of which to see is the long line of stars from the star at RA 17h 10m, Dec −15°, to the star at RA 15h 40m, Dec +7°. The latter is in fact in the Snake's Head together with the faint companion close to it and to the south east. This line points towards Arcturus and has two pairs of stars in it. Despite all these indications of the stars in Ophiuchus, you may find it difficult to find at first.

Low in the summer sky you may see a bright red star, rivalled in colour only by the red planet Mars, and this is its name, rival of Mars, Antares. At RA 16h 26.5m, Dec −26° it is not quite as far south as Fomalhaut, and is about the same magnitude, 1.22. Antares is in Scorpius, a bright and reasonably realistic constellation, which is yet another of the circle of animals. The story goes that the Scorpion bit the foot of Orion so that when they were put into the sky, the gods were considerate enough to allow Orion to hide when the Scorpion puts in an appearance, and to stay out of sight until his arch enemy has gone again. Accordingly, Orion sets as Scorpius rises, and vice versa, and we associate Orion with mid winter, and Scorpius with mid summer. The ecliptic passes through the northern part of the constellation, very low in the sky for northern observers.

The Summer Triangle

High in the sky and in the south towards the end of July at midnight is a large triangle of first magnitude stars, the Summer Triangle, shown in Fig 31. On the western point is Vega, α Lyrae, the bright star of the Lyre. Vega (RA 18h 35m, Dec 39°) has a magnitude of 0·14. A fine white star, Vega is just about circumpolar at the latitude of Greenwich.

The constellation of Lyra is quite small and with the naked eye there are only three other stars which are easily seen. A pair of third magnitude stars, γ and β, are found to the south east of Vega, and only about 7° away from it. Even closer to Vega, and to the north east is a famous fourth magnitude star, ε. With a pair of binoculars, this star is seen to be two stars. They are separated by an angle of 3′ 28″, about one tenth of the apparent diameter of the Moon, so that ε Lyrae should appear double to a very keen naked eye. So far, there is little remarkable about the star, but when examined through a fairly powerful telescope, not one but both of the two stars which form ε are seen to be double stars. Thus ε Lyrae is a double double star!

Another wide double which can be separated with binoculars, though not as easily as ε Lyrae is found in the incredible star fields of the Milky Way at the tail of Cygnus, the Swan, which is also called the Northern Cross. This is the large cross to the east of Lyra, of five main stars, and the double star which you should examine through binoculars is β Cygni, called Albireo, at the foot of the cross. This star is in the line from Vega, through γ and β Lyrae. The components of Albireo are bright blue and gold, making this star a serious contender for the title of the most beautiful star in the sky. Although really needing a telescope of moderate size to show the components of Albireo to advantage, the colours can be seen with binoculars of reasonable quality. With your binoculars on Cygnus, it is well worth sweeping through the whole region. The constellation lies on the galactic equator, in the middle of the Milky Way, and the view is breathtaking. Along the middle of the Milky Way near Albireo and the next star in the cross, η, and running parallel to both just underneath, is a curious rift in the Milky Way. This is due to dust clouds in the middle of our galaxy which obscure the stars behind.

At the northern end of the middle of the cross is the second bright star of the Summer Triangle, Deneb, α Cygni (RA 20h 40m, Dec +45°, magnitude 1·33). This star is also circumpolar from the latitude of Greenwich.

At the lower point of the Summer Triangle is Altair, in the constellation

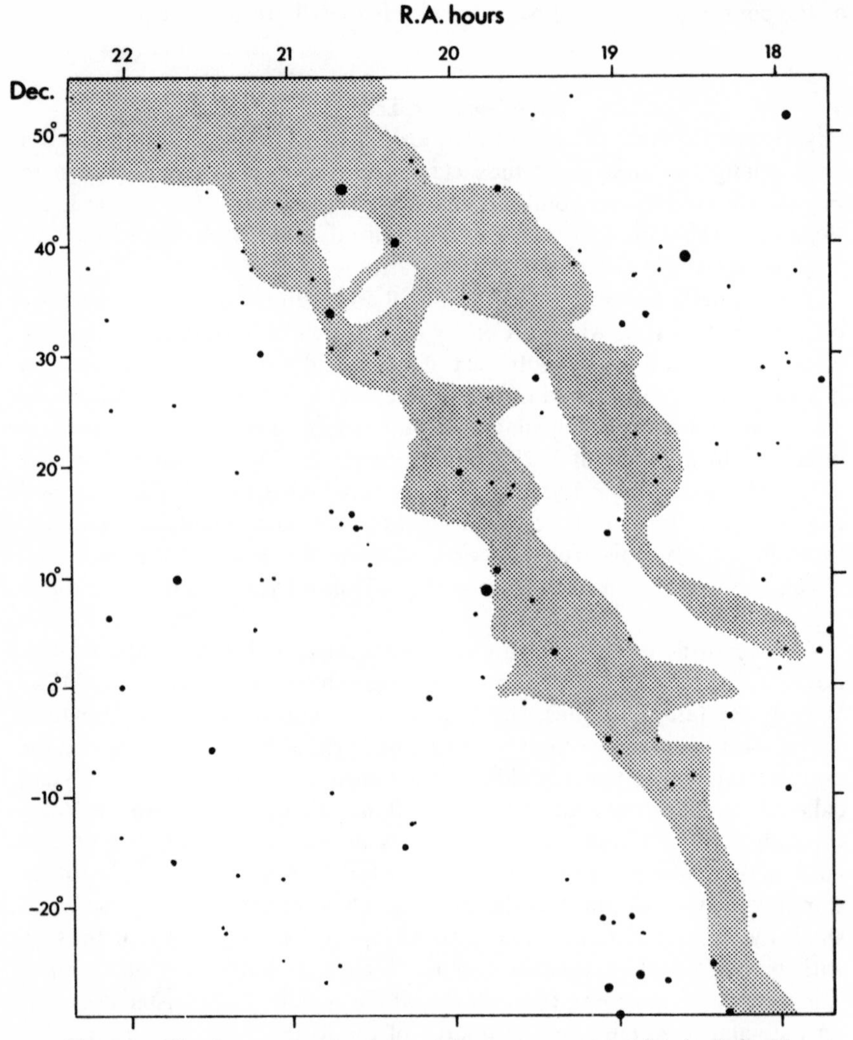

Fig 31 The late summer, the southern aspect at midnight July 22.

of Aquila, the Eagle. Altair, or *a* Aquilae, is at RA 19h 48m, Dec +8° 43′, and its magnitude is 0·89. The Eagle's tail points south west towards Scorpius, and its wings are spread either side of Altair, the most easily seen being the stars either side of Altair with which they form a distinctive trio pointing towards Vega.

Close above Aquila, and just to the east of the line from the three stars pointing to Vega is a very small constellation of four stars, all very faint,

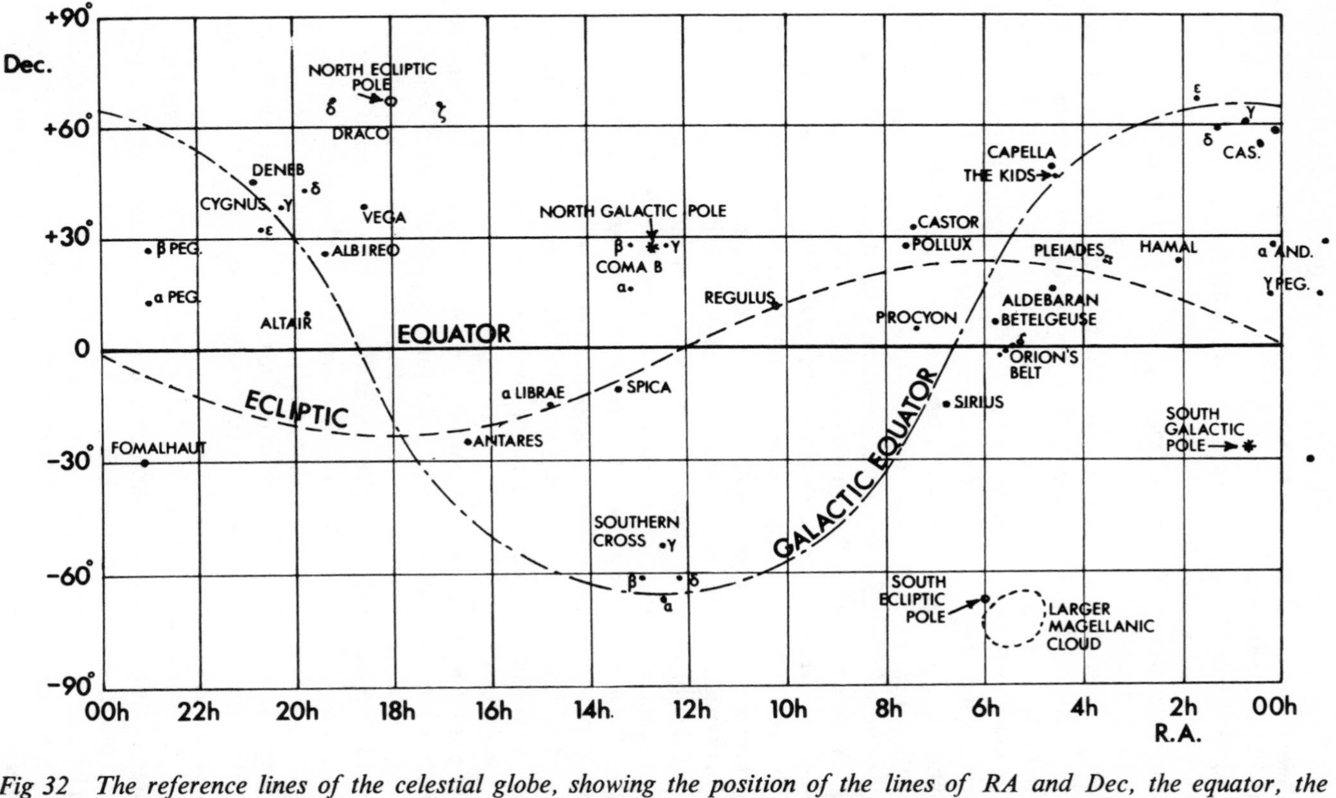

Fig 32 The reference lines of the celestial globe, showing the position of the lines of RA and Dec, the equator, the ecliptic, and the galactic equator, together with some of the more prominent objects near them in the sky.

called Sagitta, the Arrow, which is apparently flying towards the Great Square in Pegasus. Not a very brilliant constellation, but you couldn't have a much better arrow.

Another constellation which is small and faint, and yet distinctly like its name, is to be found to the north east of Altair, and not far from it and to the south of Deneb. This is the Dolphin, Delphinus, just reaching the top of a joyful leap.

To the east of Delphinus are the westerly stars of Pegasus. Beneath Delphinus and Pegasus is the zodiacal constellation of Aquarius, the Water Carrier, which contains few obvious bright star patterns. Two stars are shown to the south of the end of the Eagle's wing. The star to the north is a naked eye double, *a* Capricorni, with a separation of 6′ 16″. This double, and the star to the south, are the most easily found in the constellation, the Sea Goat, Capricorn, also in the zodiac. The most southerly of the constellations in the zodiac, Sagittarius, the Archer, is not well seen from British latitudes, although it contains a sprinkling of second and third magnitude stars which are fairly close together at about RA 19h, Dec −27°.

That concludes our quick journey around the northern skies, and only a tiny proportion of the interesting objects to be seen have been mentioned. But as explained at the beginning of the chapter, it is best to get a working knowledge of the most easily recognised parts of the sky, and to know when they can be seen. In the exercises and experiments which follow in the next chapters we shall need to refer to the stars and constellations, so it is of practical importance to become familiar with them. The background of imaginary lines of RA and Dec across the celestial sphere are, when all is said and done, still only a convenient way of fixing a position for each object.

The main indicators of the positions of the equator, ecliptic, and the galactic equator are shown in Fig 32. Features of the sky south of Dec −40° are of course not well seen by observers in northern temperate latitudes, but it is interesting to note that two of the most famous features of the southern sky, the Larger Magellanic Cloud and the Southern Cross, are both conveniently situated to indicate the south ecliptic pole and the southernmost point of the galactic equator. The poles of the ecliptic and the galactic equator are the points in the sky 90° away from the respective circles around the celestial sphere.

To learn the positions of the fainter stars and gain a great deal of practical data at the same time, you cannot do better than acquire a comprehensive star atlas (see p. 201).

The positions of the stars given in this chapter are approximate, and in most cases, based on their positions in 1950: see the table in the Appendix for star positions as in 1971.

EXERCISES ON CHAPTER 4

1 What is the minimum altitude that Deneb can have from latitude 51° north?

2 In Chapter 3 we saw how to use the Pointers, and one of the stars in Cassiopeia (which?) as clocks in the sky. Several stars make useful direction finders as well. How could you find east and west at certain times, assuming Polaris cannot be seen? How could you find south with a watch and the star Regulus (or any other star)?

3 Draw from memory the general shape of the constellation of Orion. What constellations are in the form of a cross, and a pentagon? What are the bright stars they contain? Which is the most favourable time of the year to look for both during the late evening?

4 Now read again the part of this chapter which deals with the sky visible at the moment, and study the appropriate chart, then go outside and, looking at the sky, see how much of this information you can remember and relate to the stars you can see.

5

Models of the Celestial Sphere

Most people have by now been to, or at least heard of, a planetarium. In a way, these have been named rather misleadingly since the planets form only a small part of the display. Unfortunately, 'displays' are nearly all some of the best planetaria (planetariums) are used for, and while it is praiseworthy to try and excite the passing interest of the public in the splendours of the night sky, these expensive instruments are seldom used to the limits of their abilities in demonstrating the practical aspects of the subject of astronomy.

Because a planetarium is so useful in learning the constellations, we are going to make one using little more than old newspaper and some left-over wallpapering adhesive. You will not, unfortunately, be able to see the Sun, Moon and planets on your planetarium's sky, but in the next part of this chapter we will put right this deficiency with another kind of model of the celestial sphere. Your planetarium will be used for learning the stars only, and, it must be admitted, for a little amusement also.

The planetarium takes the form of a partial sphere of papier maché, about 50cm diameter, cut around the circle of declination which is your southern limit to make what looks like an oversize space helmet, and into which you put your head to look at the stars.

The first requisite is a spherical mould, or at least hemi-spherical. The planetarium could be made on the bottom of a Space Hopper or a large beach ball would be ideal. The mould must be of the required diameter, ie about 50cm, and should be placed in a suitable position so that at least the top half is free. Placing the ball or what-have-you in a plastic washing-up bowl would do very well, particularly as the next stage tends to be very messy.

Wet the surface of the mould and cover it with patches of torn-up newspaper to a little way below the 'equator'—that is, its maximum diameter. The pieces of paper need not be very small, but the smaller you have the patience to use, the less the likelihood of causing wrinkles. It is a good idea to keep the pieces of paper very wet, and if you tear up a supply for each layer and place the pieces in a bowl of water they will be just about right. Apply two layers of wet paper without any glue. This is to ensure that the finished hemisphere does not stick fast to the mould.

Now apply a third layer, pasting each wetted patch with the wallpaper adhesive on both sides. About five or six glued layers should be applied, taking care to take each layer a little below the equator, but applying each successive layer a little less below this line, so that on the finished half there will be five or six layers down to about the equator, and then it will taper in thickness below.

You must now leave the papier maché on the mould for several days to dry out thoroughly. When there are no damp patches left, carefully free the edge by slipping a flexible knife blade underneath. The hemisphere can then be lifted off the mould, and the unglued paper pieces which were first applied can be carefully removed from inside.

Now repeat the whole process, to produce a second hemisphere. The half globes are surprisingly strong, and remain in perfect shape during the subsequent handling. The two halves can now be placed together, and the approximate position of the most southerly circle of declination visible from your latitude, 90° minus your latitude, marked on one of them. Cut around this line with a pair of scissors, but it is not necessary to take too much trouble with this cut as it can be trimmed later.

To assemble the two parts of the planetarium, tear some fairly long thin strips of newspaper and put them in water. Tear away some of the paper at the bottom of the half sphere, and around the upper edge of the other part (from which you have removed the 'pole', below the southern Dec circle). If possible, one half should be made to slip over the other at the equator. Having completed this preparation, apply plenty of water to both sides of the join on both parts. Then apply adhesive to the same surfaces and assemble the two halves. Build up the interior or the exterior surfaces of the planetarium as necessary to give as smooth a surface as possible, applying adhesive liberally all the time.

Allow the part sphere to dry thoroughly for a day or two, then apply a coat or two of acrylic white undercoat to the outside surface to cover the printing on the newspaper pieces and to provide a white, matt surface on which you can write with a ball-point pen. Now you must judge as closely as possible the top central point of the part sphere and mark a dot to represent the pole. Stand the model on a flat level surface and if the pole is not on the highest point, right in the centre, trim the bottom edge a little to square the whole thing up, but be careful not to cut too much away and lose your southernmost declination!

If you now place a square on the table and slide it up to the side of the model, it should touch it at the widest diameter, which will be the equator, and you can now either make a series of marks all the way round the equator where the square touches it, or, with a needle inserted in the pole and a piece of thread tied near the tip of a ball-point pen and to the needle, thus making a crude pair of compasses, draw the equator at the appropriate distance.

Now measure the distance around the circumference of the part sphere at the equator and from the equator to the pole with a tape measure, preferably in millimetres. Divide the circumference length by 24 so that you can mark 24 equally spaced points around the equator. These points are where the lines of RA will cross the equator. Divide the equator-to-pole distance by 9, to obtain the distance representing 10° of Dec.

Now draw the remaining circles of Dec and the hour circles of RA. Using either an extending pair of compasses or the crude string compasses centred on the pole, draw the Dec circles. This gets easier the closer you are to the pole, but near the equator, and in particular below it, it can become quite awkward. Below the equator you may have to wrap the tape measure around the sphere at the appropriate Dec and draw the circle freehand. Alternatively, use a normal pair of compasses set at the distance apart of the Dec circles and draw the −10°, −20° circles, etc by keeping the point on the equator while the line is drawn below it.

Since the hour circles each pass around the circumference, a piece of string can be tightly wrapped around, passing through the pole and opposite hour marks. The hour mark on the opposite side has a difference in RA of 12h. Draw the line freehand along the string. The finished lines may not be absolutely accurate, but they will be near enough for our purposes.

Now you can mark the stars in their right positions on the outside surface; or can you? Which way do the hours of RA go around the equator on the model? Remember that when in use you will put your head right inside the sphere, so that from the outside you are looking at the sky from the outside also. This is common to all celestial globes, which is what our planetarium is, except that it is hollow for viewing from the inside. The hours of RA must be marked anticlockwise around the pole. If the sphere is standing on the table in front of you, and you mark 00h on the hour circle immediately facing you, then 1h will be the next on the *right*. For southern observers, the 1h circle will be to the *left* of the 00h circle.

When you have marked the stars on the planetarium you will find the familiar patterns of the constellations have become mirrored. The Plough's handle will bend away to the right of the body of the Plough, for example. This means that to mark the star positions on the sphere you must measure the RA and Dec of each one from the charts or a star atlas to place it in the right position. This may seem very tedious, but you will soon become used to gauging the position of one star as the mirror image of its position in the chart relative to the adjacent stars. To help read the star positions quickly you could trace each chart and draw in the lines of RA and Dec which are marked around the borders, and possibly add the lines half way between each, or even more if you wish. You must also devise a code to show the magnitude of each star you plot on the sphere. It is little use plotting stars below magnitude 4, and it is very difficult to show intermediate magnitudes, so that it is best to use four symbols only for magnitude (you can always modify Sirius later) and choose the nearest integral magnitude for each star. Add the names and Greek letters to as many of the stars as you can. All this marking, the lines of RA and Dec, the stars and their names etc should be marked clearly, using, for example, a ball point pen. But it is advisable to mark the star positions lightly in pencil first, as inevitably one or two are at first marked as in the charts, and not in the mirror image position.

A coat of varnish over the completed outer surface can be applied now, to make cleaning easier and to protect the white surface and the star markings. When this is dry, apply a coat of matt black paint all over the inside surface and allow this to dry thoroughly as well. The very last job is to find four sharp, round pointed objects to pierce the surface at each star. This is why you need the code for magnitudes, because each star must be pierced with one of four different diameters, the first magnitudes with the largest, the second magnitudes with a smaller one, and so on. The smallest hole for the fainter stars should not be so fine that you cannot see any light through the hole when your head is inside the planetarium. In any case, a very fine needle will probably not be able to pierce the papier maché, which is amazingly strong. It is a good idea to examine the effect of three or four different magnitude stars close together at the start of this final stage, to test the result for reasonable contrast, allowing you to see the fourth magnitude stars without giving a hole for the first magnitude stars through which you can see the details of the room outside the sphere. Orion is a very good constellation on which to do this test.

The paper planetarium cannot be used for the spectacular demonstrations which take place in a planetarium with an optical instrument, but it is ideal for learning your way about the sky at your fireside. By moving a finger over the outside of the sphere while your head is inside, you can make any star you wish to identify go out and, keeping your finger in position, then look at the outside to see which star your finger is covering.

The Celestial Globe

The celestial globe is just about the most informative and versatile calculating and demonstration device you can make. It is a 'must' for any reader of this book.

The celestial globe is a working model of the sky which can be set to show the sky at the observer's latitude, or any other, including the northern and southern skies. For this reason, celestial globes were once used by sailors for navigational purposes. It can be made as simple as you like: a globe with lines of RA and Dec will do, or you can mark all the stars on its surface and have movable planets as well. It depends on how much time you can afford to spend on it.

To make a simple globe, the biggest problem is to find a suitable sphere to use. Size is not very important but, in general the larger the better. The obvious choice is a rubber ball, but if this is of the air-filled variety as most larger diameter balls are, your painstakingly made globe will slowly deflate, breaking free of its mounting in the process. Solid rubber would be better, but solid rubber balls are usually rather small. Wooden balls are also suitable, but expensive. Several proprietary games are sold which include fairly large and solid balls, and with any luck you may find a slightly the worse for wear set in a sale from which you can appropriate a suitable sphere.

If, while you are at the sale, you find a terrestrial globe in reasonable condition, your problems are over. To use a terrestrial globe already mounted on an axis of rotation involves a different technique to the model made from an ordinary ball, so we will consider these separately, and start with the unmounted ball.

The first job is to measure the diameter of the ball by placing it on a table and sliding two squares up to it so that they just touch on either side. Then measure the distance between them. Cut out an open rectangle of fairly thick card with an inside width a little greater than d, the diameter of the ball, the outside width being sufficient to give it some rigidity. The two uprights should be at least 12mm longer than the radius of the ball, ie $d/2$, as shown in Fig 33a. Stick a piece of card at the tops of the upright arms and make a mark on each at a height of exactly $d/2$ from the inside bottom edge.

Insert a drawing pin in each side at the marks, with the heads on the inside. Now apply a strong adhesive to the tops of the drawing pins, which are facing each other in the jig and, if necessary, to the ball also at diametrically opposite points, the poles. Place the ball in the jig so that the poles are between the drawing pin heads. When the glue has set, remove the ball (you may have to destroy the jig) and it will have two points sticking out on either side, as extensions of its polar axis.

Now draw two concentric circles on card, the inner one equal to the diameter of the ball plus about 2mm, so that when this circle is cut out the ball will be free to rotate inside it. The outer circle should be about 15 to 20mm greater in *radius* than the inner circle. Draw a common diameter across the circles, then mark the inside circle all the way round at 10° intervals, starting at one of the ends of the diametrical line. Cut out the inner circle, and cut round the outer, to give a ring of card. Now cut out two squares of card, the same width as this ring. These will be stuck over the marks to hold the axis of the ball and must not protrude either inside or outside the ring. We will call this ring the meridian circle.

Now draw another pair of concentric circles on card. The radius of the inner circle should be *less* than the outside radius of the meridian circle by about 10mm, and its outside radius should be at least 25mm greater than its inner radius. Draw a common diameter across both circles. With your compasses at the common centre of the circles, and set to about $\frac{1}{2}$mm greater than the outside radius of the meridian circle, strike an arc across both ends of the diameter you have just drawn. With your protractor,

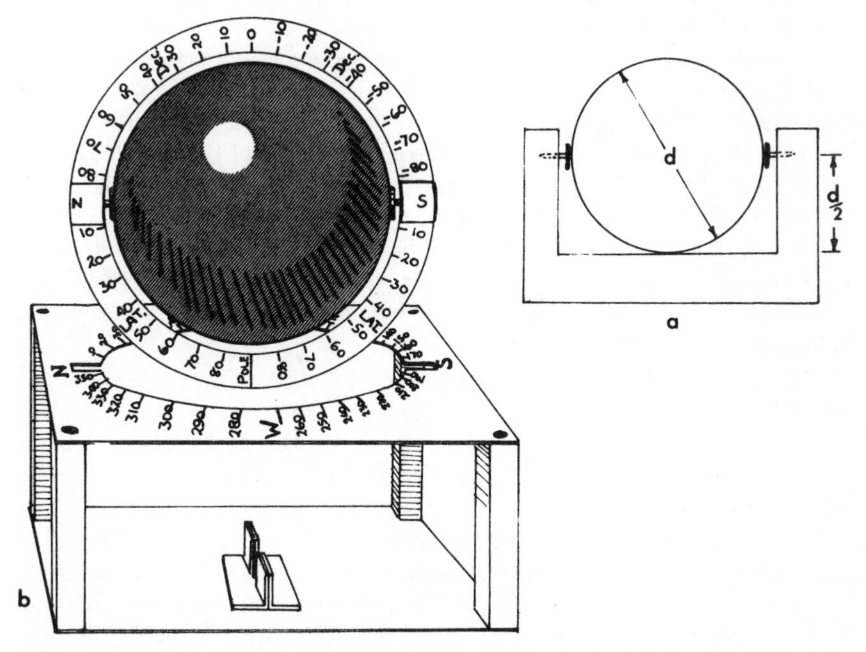

Fig 33 (a) *A jig is used to fasten drawing pins on the 'poles' of the ball, so that it can be mounted in the meridian circle as in* (b), *and the whole assembly is supported in the azimuth ring, with a guide in the base of the mounting.*

mark around the inside circle at 10° intervals starting from the common diameter line. Now cut out this second ring, which we will call the azimuth ring. Cut a slot the thickness of the card on opposite sides along the diametrical lines from the inside of the ring up to the limit marked by the arcs. You should now be able to place the meridian circle inside the azimuth circle and at right angles to it, through the slots.

Cut four pieces of wood, ensuring that the ends are square, with a length of about 5mm longer than the outside radius of the meridian circle to act as legs for the azimuth circle. The azimuth circle must be fastened to these legs, which should be spaced at 90° intervals around the circle, and at some 45° to the slots into which the meridian circle must fit, as can be seen in Fig 33b. The legs can be fixed with more drawing pins, or glue. The bottoms of the legs should be glued (or pinned) to a base, the size, shape, and material of which is your choice. Now when the meridian circle is placed in the slots so that it is vertical, and with its diameter marks level with the azimuth circle, it will be about 5mm above the base at the lowest point. Glue two L-shaped pieces of card back to back, cut a small slot in the top down to this clearance above the base, and glue in position on the base so that it acts as a support for the bottom of the meridian circle, and keeps it vertical.

Using the azimuth circle as a table, place the meridian ring flat upon it and place the ball on the meridian ring so that the drawing pin points are exactly on the diameter marks, and the ball is central in the ring. Then glue small securing pieces in position over the points of the pins. When the glue is set, the ball will rotate freely inside the meridian ring, and the whole assembly can be placed in the azimuth ring, with the bottom of the meridian ring in the guide slot on the base. The inclination of the sphere's axis can be adjusted through 180° (except where the axis supports will not pass through the guide slot in the azimuth ring; but this is a small limitation) to simulate various latitudes.

The lines of RA and Dec should now be marked on the ball. These should be marked using the model itself to guide the pen, and using the 10° marks, as follows. Place the pen point on the ball at the equator, that is half way between the poles on the meridian ring. Holding the pen firmly against the meridian ring, turn the ball on its axis, and the equator will be drawn. Draw the other Dec circles in a similar manner, with the pen at each of the 10° marks in turn. It is probably easier to draw these circles with the meridian ring turned so that the axis of the ball is nearly horizontal.

Now mark the equator at 15° intervals. This could possibly be done using the 10° intervals on the azimuth ring, with the axis of the ball vertical, but since the guide on the base will usually prevent the axis supports from assuming this position, it is best to measure the equatorial circumference of the ball and divide by 24, as for the marking of the planetarium. The hour circles of RA can be drawn using the meridian ring as the guide for the pen, setting each of the 24 marks on the meridian ring in turn, and drawing half of each hour circle from pole to pole. Number these circles from 0 to 23h anticlockwise around the north pole.

You must now draw the ecliptic on the sphere. This is, fortunately, a great circle, which means that its centre is at the centre of the sphere, and it can be drawn using the azimuth circle as a guide. All you need to do is mark the most northerly and southerly points of the ecliptic at RA 6h, Dec $+23\frac{1}{2}°$, and RA 18h, Dec $-23\frac{1}{2}°$. Now place the ball in the stand, and incline it in the meridian ring so that these two extreme points on the ecliptic and the two equinoxes (00h and 12h, on the equator) are all just level with the top surface of the azimuth ring; then draw the ecliptic, preferably using a dotted line to distinguish it at a glance from the lines of declination, using the azimuth ring as your guide, but not rotating the ball. You can now mark the principal star positions on the sphere, remembering that this must be done by using the co-ordinates of RA and Dec because, like the planetarium, the globe shows space from the 'outside'.

With this model of the celestial sphere you should mark in the values of declination on the meridian ring, although there is nothing to stop you marking them on the ball itself if you feel inclined. We will leave the description of the use of this model until later, after describing the making of a celestial globe from a globe of the Earth.

If you find a globe of the Earth in reasonable mechanical state (the condition of the map is immaterial) at a sale, you can make an elegant celestial globe fit to grace your drawing room, which will be an endless source of quick reference.

A metal globe is best for its ease of marking, and also for the possibility of putting magnetic planets, Moon and Sun on its surface. Usually, an Earth globe has a fixed latitude scale which is a ready-made meridian ring and declination scale for the celestial globe. The globe will be on a stand, and inclined at $23\frac{1}{2}°$, which is a minor inconvenience. You can either take the globe off the existing stand, and make a new stand for it in exactly the same way as described for the model mentioned above, or make a horizon which can be inclined to suit the latitude instead of turning the whole globe in its mounting.

There are a number of ways in which the Earth globe can be converted, but for the benefit of those who are not confident in their do-it-yourself abilities the following procedure works well.

Measure the diameter of the globe with calipers, or by measuring the circumference with a piece of string and dividing this by π (3·142). The

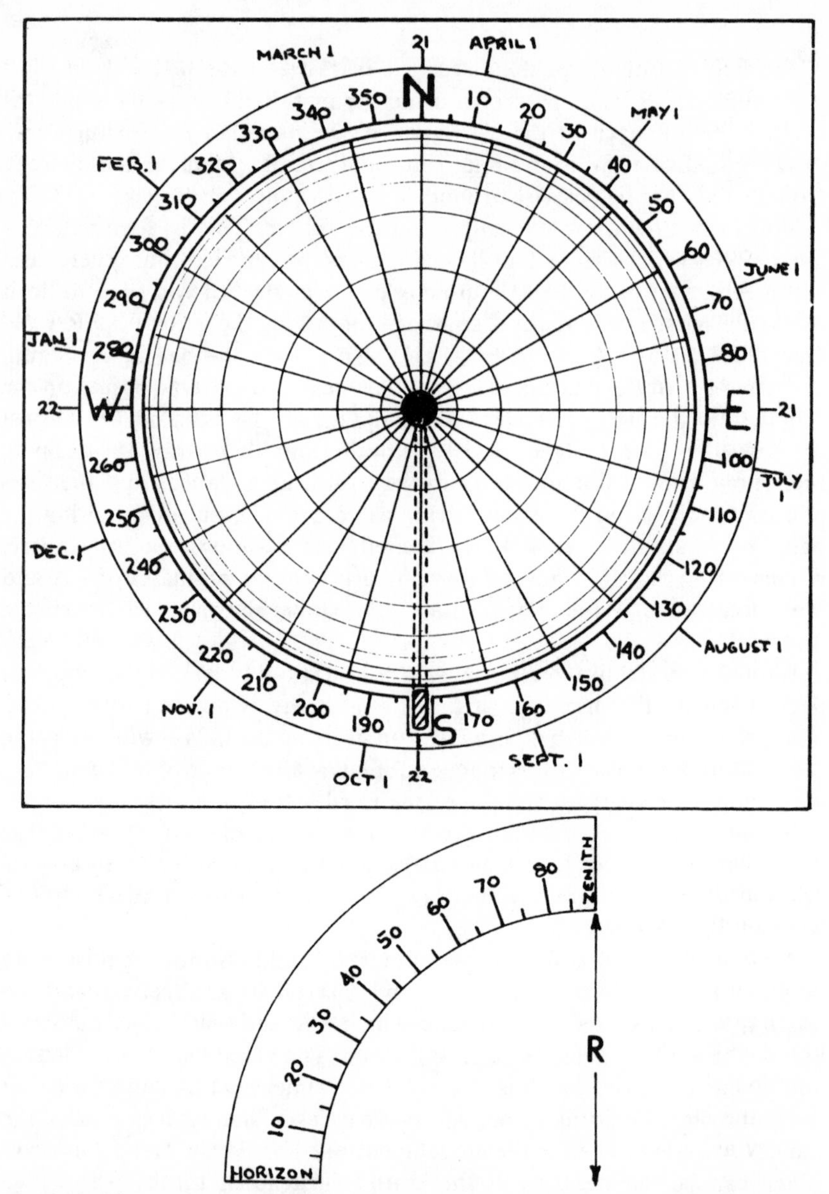

Fig 34 The azimuth circle is calibrated in degrees of azimuth and the dates of the year. The meridian ring is made to pass inside the horizon ring and is shown in the diagram as a cross section where it passes through the ring (also shown as a broken line to the pole). A simple quarter circle of radius R equal to that of the sphere is used to measure altitude. It can also be used to measure angular separation on the sphere.

horizon circle will usually pass over the meridian ring (latitude scale). Cut a square of thin plywood, about 4mm thick is sufficient, the sides of the square being the diameter of the globe plus *twice* the distance of the outside edge of the latitude circle to the surface of the sphere, plus about 20mm. Find the centre of the square by the point at which the diagonals intersect, and draw a circle at this centre with a diameter equal to that of the globe plus 6mm, which will allow 3mm clearance all round. This circle should now be divided into 360° at 5° intervals, with the marks made plainly visible on the outside of the circumference. The first mark should be made exactly half way along one side of the square.

Now carefully cut out the circle with a fretsaw, bearing in mind that it is the outside piece which you are going to use! A slot must now be cut for the meridian ring from the inside of the circle. The slot should be cut at the 180° mark of the 360° scale, so that when the horizon is mounted on the model, the slot will be due south.

Cover the globe in black paint, and when the paint is dry, place the horizon in position around the equator so that it is exactly half way down the sphere. Measure the distance from the westerly point, or the easterly point, to the base. Now you must make a support for the horizon so that it is held at this half way position, and is free to rotate about the east–west axis. You may care to devise your own supports, but a very simple method is to use two straight pieces of thick steel wire of the type strung between posts for chain link fencing. The bottom of each piece can be bent so that it rests horizontally on the base of the globe, and a small loop made in the end so that it can be bolted to the base. About 12mm at the other end of each piece of wire is also bent horizontally so that the distance between the two bends in each piece is the required height of the horizon from the base.

Two small pieces of the same wood from which the horizon is made are then glued to the underside of the horizon on the extreme edge of the square, under the east-west centre line. The extra thickness of the horizon so provided allows a hole to be drilled towards the centre of the circle in a horizontal direction. The diameter of the holes, which must be drilled on both the east and west points, must match the diameter of the wire as closely as possible to obtain a tight fit on the supports. The horizon can then be tilted to the desired latitude and will remain in this position.

The markings of the horizon circle are shown in Fig 34. The azimuth scale starts at 0° at the north point, opposite the slot for the meridian circle, and is marked in 5° or 10° intervals, depending upon how much room you have, through east (90°), south (180°), and west (270°), ie clockwise. You can now mark a date scale around the circle. This will allow you to set the globe by mean time as well as sidereal time and, incidentally, provide you with another mean time-to-sidereal time converter. The dates will also be marked clockwise, and the autumn equinox will be due south on the azimuth scale, the winter solstice due west, the spring equinox due north, and the summer solstice due east, with the intermediate dates as shown in Fig 34. The dates can be further subdivided as required. Remember that at longitudes other than that of the central longitude of

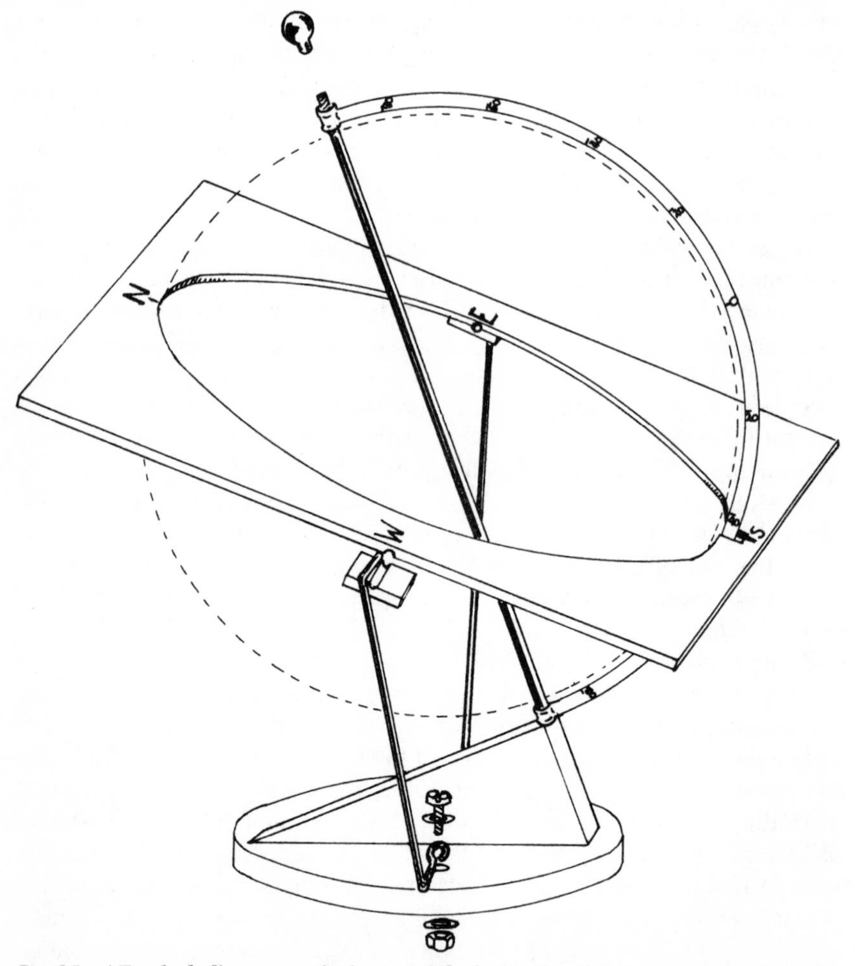

Fig 35 'Exploded' view of the modified Earth globe mounting for the celestial globe, with the sphere itself shown as a dotted line. Note that the original latitude scale becomes the Dec scale, which can also be used to set the latitude of the horizon ring.

your time zone, the setting of the sphere must be made 4 minutes fast or slow for each degree east or west. To enable altitudes to be measured, cut out a quarter circle with an inside radius equal to the radius of the sphere, and mark it in degrees from 0° to 90° (0° on the horizon). This is then placed with one end on the horizon, the other at the zenith, in such a position that it passes through the object of interest; the altitude is read off the scale and the azimuth can be read off the scale on the horizon circle.

The globe is assembled as indicated in the 'exploded' diagram, Fig 35. The 15° intervals for the hour circles can be marked around the globe's equator from the azimuth scale. The circles themselves can be drawn on the sphere with a fine sharp pointed instrument using the meridian ring as your circular ruler to guide the point, which scratches the circles through the black paint. Now by holding the point of your 'scriber' at the 10° mark on the meridian circle and turning the sphere on its axis with the point on its surface, you will mark out the declination circle for 10°, and so on for all the northerly and southerly declinations.

Now set the globe for sidereal time 18h, that is with the 18h circle on the meridian, due south, and set the latitude of the horizon circle to 23½° north, at which setting the horizon circle will be truly horizontal. The ecliptic now coincides with the horizon, and the globe is set for the sky as seen at the Arctic Circle. With the globe set to this position, and holding the horizon circle firmly at that latitude, you can now mark the ecliptic on the surface of the sphere, preferably with a fine dotted line, using the horizon as a guide. Note that the sphere must not be allowed to move from the 18h sidereal time position while you mark the ecliptic; you cannot mark this circle by rotating the sphere as you did for the declination circles.

Using a fine pointed brush and some quick-drying white paint, such as an acrylic undercoat, mark the stars on the sphere at the proper RA and Dec positions, remembering that the globe's surface as we look at it is a mirror image of the patterns we see in the sky. Make the spots of different sizes to correspond with the different magnitudes (to the nearest whole number). Now we are ready for the Sun, Moon, and planets. The big advantage of an Earth globe for making a celestial globe is that metal globes can be obtained and the movable objects in the sky can be made of magnets so that they can be placed on the globe in any desired position.

A useful magnetic material to use is Feroba. This is a special thin rubber sheet, to which a barium ferrite has been added to make it magnetisable. One surface of the sheet is a north magnetic pole, the other a south pole, so that you can cut out a small magnetic disc with an ordinary pair of scissors. Feroba can be purchased in small quantities from Magnetic

Rubber Ltd, Soho Street, Sheffield, S11 8HB, or one of this company's distributors. In the United States, these are James Neill & Co (USA) Ltd, British Distribution Services Ltd, 403 Kennedy Boulevard, Somerdale, NJ08083.

You should mark the magnets for the Sun, Moon and planets with a quickly distinguishable symbol and colour; for example Mars could be a small red circle, Venus a white crescent, and so on. Obviously the size of these symbols is out of all proportion to the rest of the sky, but this doesn't matter particularly. We must remember to make any measurements involving these objects to the centre of the symbol. A possible solution to the movable objects problem with a plastic globe is to cut tiny round pieces from a polythene bag on which you have painted your symbols for each planet. You could even try very small rubber suckers. These could be cut off those soap holders which are now made consisting of a flat disc with small suckers on both sides.

Using the celestial globe

The uses of the celestial globe are almost unlimited as far as positional problems are concerned. If you set the sidereal time for a particular moment on the globe by putting the appropriate hour circle on the meridian, or set the mean time as indicated by the hour circle, setting it opposite the date on the horizon, you have a model of the sky for that moment. If you now point the axis of the globe to the celestial pole in your sky, an imaginary line from the centre of the sphere to the sky through any of the objects on the globe will point to the actual object in the sky. So if we see a bright star through a gap in the clouds one evening, we need only note its position carefully, then set the globe to the time of the observation and align it with the pole to see which star on the globe is in the position of the star observed.

Celestial globes were once used extensively for navigation. To illustrate how, suppose that you are on a ship and the Sun is observed due east at an altitude of 50°. The date is June 1, and the time on our chronometer is 13h UT. Set the 12h hour circle to June 1 with the horizon level with the celestial equator, ie as seen from the north pole, so that we can read the hour circles against the date. Now at 12h UT the Sun will be due south, so we place the Sun on the globe under the meridian ring, on the ecliptic, and find that it is at RA 4h 30m, Dec +22°. At the time on our chronometer, 13h, the Sun will be one hour past the meridian, so that the sidereal time will be 4h 30m + 1h = 5h 30m. Now the Sun was observed due east, so it is clearly before local midday where we made the observation. The Sun will get higher in the sky, so that if it is already 50°, we must be south and west of Greenwich. Turn the sphere until the Sun is in the general direction of east with the horizon set for a more southerly latitude. Check its altitude with the measuring scale and if it is still too low, go further south. Try latitude 30° north. The altitude of the Sun when in the east is 50°. So this is our latitude. The local sidereal time is read on the meridian, and is 1h 30m, which is 4 hours slow on Greenwich sidereal time. Each hour slow is 15° west of the Greenwich meridian, so our longitude is 60° west.

This is, admittedly, rather crude navigation, especially on our non-precision globes, but the principles are the same when using mathematical methods to find positions on the Earth. We will return to this problem in a later chapter.

On what date does the Sun set at 19h 30m from latitude 51°N? This can be solved by trial and error methods using the globe. The Sun will set 7h 30m after noon on two dates at approximately the same interval after the spring equinox and before the autumn equinox. We must turn the globe until there is 7h 30m difference between the RA of the elciptic where it cuts the western horizon and the RA on the meridian. This represents the Sun's hour angle at sunset. We find that the RA of the Sun on one of the two dates is 9h 15m. Put the Sun in this position, and find the date opposite the 12h hour circle when the Sun is on the meridian and it is August 12. Set to latitude 51°, and turn the globe so that the Sun is on the western horizon. Now if we read the mean solar time of sunset for August 12 we find it is 19h 30m.

What about the other date? A quick glance at the globe shows that it is the Sun's declination which determines the time of setting at any given date. On August 12 it was +15°. It was also at this declination at RA 2h 45m. By placing the Sun in this position, and setting it on the meridian, we find that the date opposite the 12h hour circle is May 3, which is the other date at which the Sun sets at 19h 30m. On both these dates, the Sun's declination is the same; so is its noon altitude, and so are the azimuths of the points of rising and setting.

At some times of the year, the so-called 'new' Moon (which is really a day or two old) is seen 'on its back' from temperate latitudes, and at other times the crescent appears to be almost vertical. Why is this? With the globe it is easy to show the reasons. Set the globe to about latitude 50°, and put the Sun on the ecliptic, close to the spring equinox, say at RA 0h 30m. Place the Moon on the ecliptic at the two day old position, about 24° along the ecliptic, at RA 2h 0m, for example. Now slowly turn the sphere to the west, until the Sun is just about to set. You can see that the ecliptic cuts the horizon at a steep angle, and the Moon's illuminated portion, which will be a roughly semicircular crescent, the middle of which will be closest to the Sun, will be almost on its back. When you allow for the fact that on some occasions the Moon could be up to 5° above the ecliptic, because of the inclination of its orbit relative to that of the Earth, its position on this occasion could be 5° further north (measured at right angles to the line of the ecliptic). This would put it at about RA 1h 45m, Dec +16°. The Moon is now even closer to being vertically above the Sun as the Sun sets.

Now put the Sun just before the autumn equinox, say at RA 11h, on the ecliptic as ever, and put the two day Moon 24° around the ecliptic to the east or, even more dramatically, 5° below the ecliptic, and you will find that the Sun and Moon set almost at the same time. At the time of setting, therefore, the Sun–Moon line is virtually horizontal, and the Moon's crescent must be upright. You can also see, by turning the sphere westwards until the Sun and Moon have risen once more, that the reverse positions occur in the mornings at both equinoxes.

The more obvious uses of the globe hardly need description. You can use it to show the state of the sky at any date and time at any point on Earth. The times of rising and setting of any object can be worked out very simply. If you keep the positions of the Sun, Moon, and planets up to date, their eastwards motion across the sky becomes obvious. The globe will show you not only when that most elusive of the easily visible planets, Mercury, is best placed for observation, but also where to look.

It is an ideal instrument to demonstrate many of the practical phenomena we are considering in this book. Here is another example. Turn the globe to show Orion's Belt just after rising and just before setting. You can see that just after rising, the Belt stars as seen in the northern hemisphere are almost vertical, and a few degrees south of due east. Just before they set, on the other hand, they are horizontal, and the same few degrees south of west. The Belt stars can thus be used to find east or west simply by noting whether they appear vertical or horizontal when near the horizon.

The remainder of the examples you should now try out for yourself, from the exercises at the end of this chapter. You will learn a lot by experimenting with the globe, making up your own problems, and by checking the answers to some of the problems set in other chapters with the globe.

EXERCISES ON CHAPTER 5

1 What time is sunrise on July 14 from latitude 51°N?
2 From a glimpse of the Moon through the clouds on the night of January 5 you judge it to be 6 days old, and its bearing was due south. What was the

time of the observation in UT if your longitude was 5° west, and latitude 51°N?

3 On holiday with your watch set to Universal Time, you measure the position of the Sun on August 1 at 13h 20m as: altitude 50°, azimuth 250°. Where are you, approximately?
4 At what latitude does Regulus set in the north west?
5 On what date is Antares due south at sunrise from latitude 30° north?
6 What is the angular separation of Altair and Vega?
7 Fig 8 of Chapter one shows the appearance of the Sickle, Leo's head, when rising at different latitudes. Demonstrate these diagrams with the globe.
8 What is the nearest to south that the Sun can rise from latitude 51°N? What is the date when this occurs?
9 On 1972 April 3, Saturn was at RA 4h 05m, Dec +19°. How long after the Sun did it set as seen from London? Should it have been visible to the naked eye?
10 To what latitude would you have to go to see the midnight Sun on August 10?

6
Measuring the Sky

A star or other astronomical object can be fixed in terms of its right ascension and declination, or altitude and azimuth at a given moment. An angle is subtended to the eye by any two stars, and just as we can plot a plane triangle once we know the lengths of three sides, so we can fix the *relative* position of, say, a planet to two stars once we know the angles between each of the three bodies.

The Cross Staff

To measure such angles in the sky we can make a simple, highly effective and useful instrument, the cross staff, another ancient navigational device. We need a piece of D section wooden beading 60cm long and a small second piece of the same beading 100mm long, a fairly short and preferably thick elastic band, and four pins with large heads (if you can get the type of pins with coloured plastic heads, white are ideal). The pins must be pushed into the centre line of the short piece of wood on the curved side. One pin is pushed in close to one end at an angle, so that the head is as nearly as possible over the end of the piece, and another pin is inserted in the other end in the same way, so that the centres of the heads are exactly 100mm apart and at the same height. A third pin is inserted exactly half way between the other pins. The last pin is placed exactly half way between one of the outermost pins and the central pin. The cross arm is then placed at right angles to the main piece, with the flat sides together, and secured with the elastic band as shown in Fig 36a. This arrangement allows the cross arm to be slid from one end of the staff to the other, keeping the cross arm all the time at right angles to the staff.

When the instrument is held up to the eye as shown in Fig 37 the outer pins make an angle to the eye of $2a°$ when exactly covering the two stars S_1 and S_2. To find the angle a, we bisect the angle (as the body of the staff does) and each half of the cross arm is then of length l, and the centre line, where the pins are, is at a distance x from the eye measured along the staff.

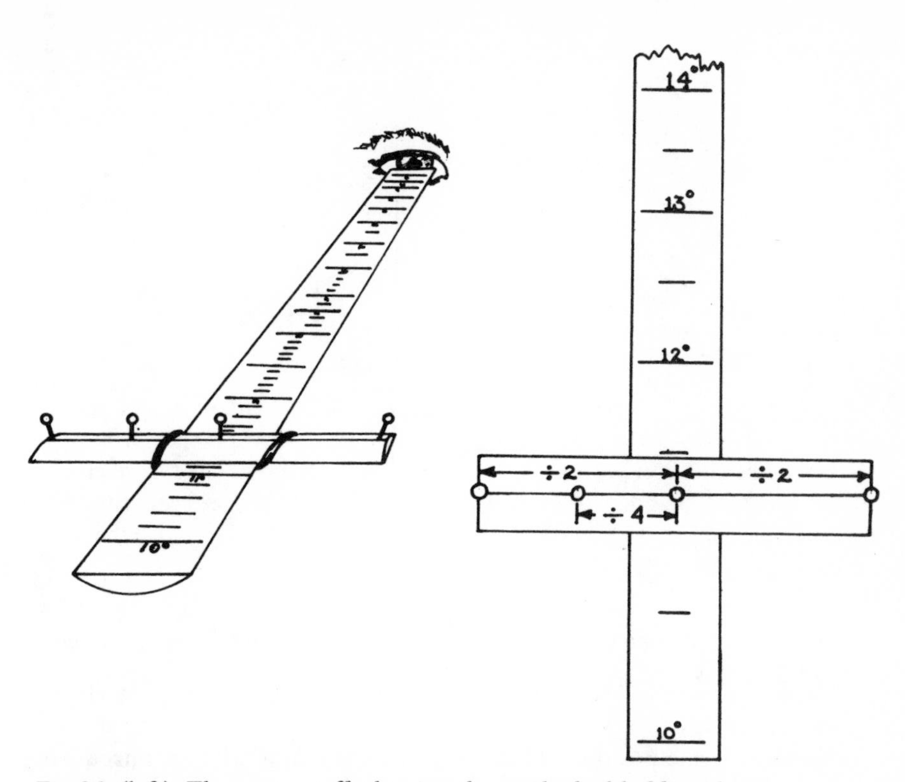

Fig 36 (left) *The cross staff, showing the method of holding the cross arm on the instrument with an elastic band.* (right) *Part of the scale at the end of the cross staff, also showing the position of the sighting pins and the corresponding scale correction factors.*

By elementary trigonometry we can say that

$$l/x = \text{Tan } a$$

$$\text{so that } x = l/\text{Tan } a$$

Since l is half of 100mm we know that $x = 50/\text{Tan } a$. We therefore mark the scale on the staff by taking an angle, say 10°, halving it, and calculating the distance $x = 50/\text{Tan } 5°$, or more conveniently if we have some cotangent tables, $x = 50 \text{ Cot } 5°$, which is near enough to 572mm. We mark a line across the staff at 572mm from the eye end, and *mark it* '10°'. We then take the cotangent of 5·5°, repeat the calculation, draw another line and mark it '11°', and so on. We can make as many scale marks as we like, but as we get closer to the eye and the marks become far too close together to mark at every 1° interval. To save the long and not very edifying calculation, the table gives results worked out at suitable intervals. You

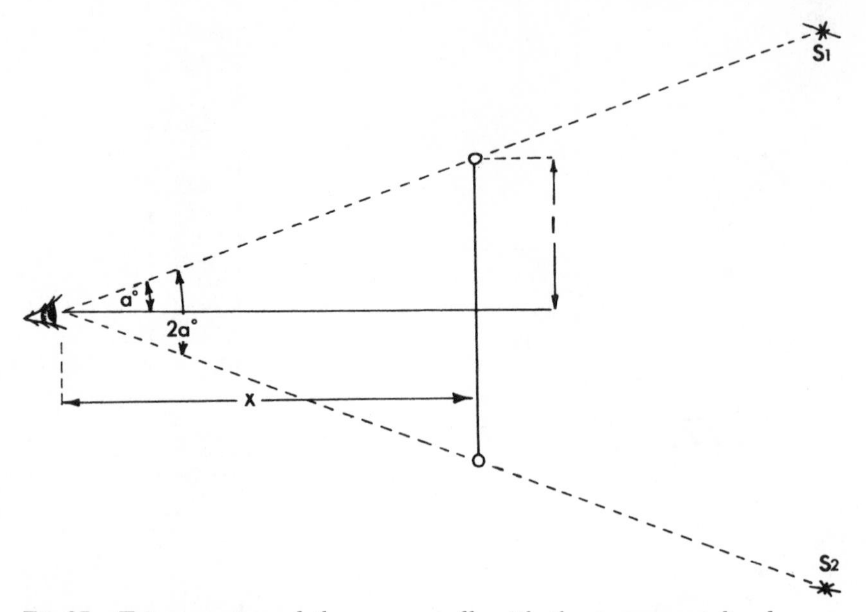

Fig 37 Trigonometry of the cross staff, with the instrument lined up on stars S_1 and S_2.

can estimate the positions of the intermediate marks without introducing noticeable errors (see Fig 36b).

The outer pins can be used to measure angles down to 10° directly. To measure smaller angles we use one of the outer pins and the middle pin, remembering to divide the scale reading by 2. Below 5°, we use the centre pin and the fourth one (the nearest to it) and divide the scale reading by 4. This last angle is not strictly a quarter of the angle between the outers, especially when near the eye, but it is near enough. (For a little practice, work out why this is not completely accurate!)

To use the cross staff, hold the end calibrated 140° immediately below the eye and lightly touching the cheek. Don't walk around with it in this position on a dark night—if the other end hits something in the dark, the consequences could be very serious. Point the staff between the objects being studied, and with the other hand slide the cross arm along the staff until the two pin heads being used lie as exactly as possible on top of the stars, planets or whatever. Because the pins are very near your eye, even at the end of the staff, they will be out of focus while you are looking at the stars, and in effect you will be able to see the stars through them. Try to keep the cross arm at right angles to the staff and centred properly. This is quite easy even with the rather crude elastic band holding method. If you

$a°$	*Mark the scale*°	$x = 50 \cot a (mm)$
5	10	572
5·5	11	518
6	12	476
6·5	13	438
7	14	407
7·5	15	380
8	16	356
8·5	17	334
9	18	316
9·5	19	299
10	20	284
11	22	257
12	24	235
13	26	217
14	28	201
15	30	186·5
16	32	174·5
17	34	163·5
18	36	154
19	38	145
20	40	137
25	50	107·5
30	60	87
35	70	71·5
40	80	59·5
45	90	50
50	100	42
55	110	35
60	120	29
65	130	23
70	140	18

want to go to a lot more trouble, you can secure the cross arm with a sliding dovetail or similar method, but this is not really necessary.

A number of interesting measurements can be made with this instrument, and the following will give you some practice and teach you some other important things in the process. By now you must be familiar with all, or at least most of the major constellations. What is one of the most striking that will be high in the evening sky today? From memory draw the main pattern of the constellation as near as you can to scale (eg 1cm per degree between stars). Now draw the constellation again while actually looking at it, still using the same scale. Now use the cross staff to measure the angles between the stars, taking care to include each star in at least two measurements. You can now draw the whole constellation from your results to the same scale as before.

By comparing the three drawings you will convince yourself that astronomical observations must never be based on memory, and even when drawings are made freehand while observing they can only give at best a close approximation and impression of what is observed.

The Quadrant

The quadrant is used to measure altitude and, with a small additional device, can be used to measure both altitude and azimuth. It is a very ancient instrument, the forerunner of the modern sextant. As we saw in Chapter five, if we know the altitude and azimuth of an object at a given time, we can fix its position on the celestial sphere in terms of RA and Dec by using the globe. In the next chapter we shall see how we can convert altitude and azimuth to RA and Dec with some precision, and if you have a good quadrant and a good watch you could use them for some surprisingly accurate navigation.

The method described here is only one of many suitable ways of making a quadrant, and so by all means make your own refinements to suit the materials you have available. The most important item is a protractor—the biggest you can find. You need only a 0 to 90° scale which is half the normal semicircular protractor, and it is possibly best, therefore, to make your own scale. It must be as large as you can make it, and still be able to handle with comfort. It is to be marked in one degree intervals, if possible with each degree being subdivided into halves, or even quarters.

Since many protractors are not as accurate as one would wish, and certainly not as big, you can mark or measure angles with great precision using the chords of the angles. The method will be useful in the remainder of this and the following chapters where angular measurements should be made to fractions of a degree in many cases. The chord of an angle is the distance between the ends of two lines of unit length separated by the angle. Its value can be obtained direct from a table, if you have one, or from the expression

$$Chord\ x = 2\ Sin\ \tfrac{1}{2}x$$

To use the chord to draw an angle, such as the divisions on the quadrant scale, set a pair of compasses with a very sharp point at a radius large enough to suit your particular task, but choose an easy value such as 10cm or 10in. Strike an arc from the point at which the angle is to be made—the centre of the quadrant's quarter circle in this case—across the line with which the angle is to be made. On the quadrant scale, 0° will be a line which will be the vertical side of the quarter circle, so make a faint mark across this line at, say, 10in from the centre. From the same centre make a second arc of the same radius at about the position of the 10° mark on the scale, for example. Now set the compasses to the radius multiplied by the chord of 10°, in this example 10 × 0·1743 (or 20 Sin 5°) = 1·743cm. With the centre at the intersection of the first arc and the 0° line, strike the 1·743cm arc to intersect with the 10cm radius arc you drew from the centre. Where they cross is the 10° point on the scale.

To measure the angle between two lines, simply strike the 10cm arc from the point where they meet, drawing the arc across both, and measure the length of this chord with your compass, dividing the result by 10 and look up the angle in the table.

Those who find this added slog just too much, don't worry, the protractor will give a reasonable answer in most cases. Mark the quarter circle with the most clearly subdivided scale you can manage. You must cut the quarter circle out with a margin around the straight edges, leaving the lines from the centre to the 0° and 90° marks, and of course the centre itself, on the scale.

The quarter circle must be pinned or glued to a straight piece of wood which is as long as you can comfortably handle. The line joining the 0° mark to the centre of the quarter circle (the 'origin') must be at right angles to the piece of wood. Push two pins into the top of the piece of wood, one about 5cm from the eye end, and the second at the other end of this length, taking care that the heads of the pins lie on the centre line and that they are at the same height. The line joining the pin heads must be parallel to the 90°—origin line on the quadrant. Insert a drawing pin in the origin, and tie a fine thread to it so that it is quite free to swing on the pin. The thread should be a little longer than the radius of the quadrant itself, and a small weight, such as a steel nut, is attached to the free end.

Instead of pins, the sights can be made as shown on the mounted quadrant in Fig 38. The eye end sight has a small hole in a right angled bracket, and the flat sight at the other end has a slightly larger hole, but with the centre at the same height as the eye end sight. This type of sight is much easier to use for measuring the altitudes of the Sun, when shadows must be used, or of very faint objects. Altitude measurements with the quadrant are not affected by the angle of the ground on which the observer stands. You can check the accuracy of the alignment of the sights with the quadrant scale using a spirit level by setting the instrument so that it reads 0°, and putting the spirit level on top.

A method of combining the quadrant with an azimuth measuring scale is illustrated in Fig 38. This is only one way of many which you could design for yourself. The advantage of using a stand is that if the quadrant is hand held, you must be very careful to hold it upright while you take your altitude reading, so that the plumb line does not touch the scale card

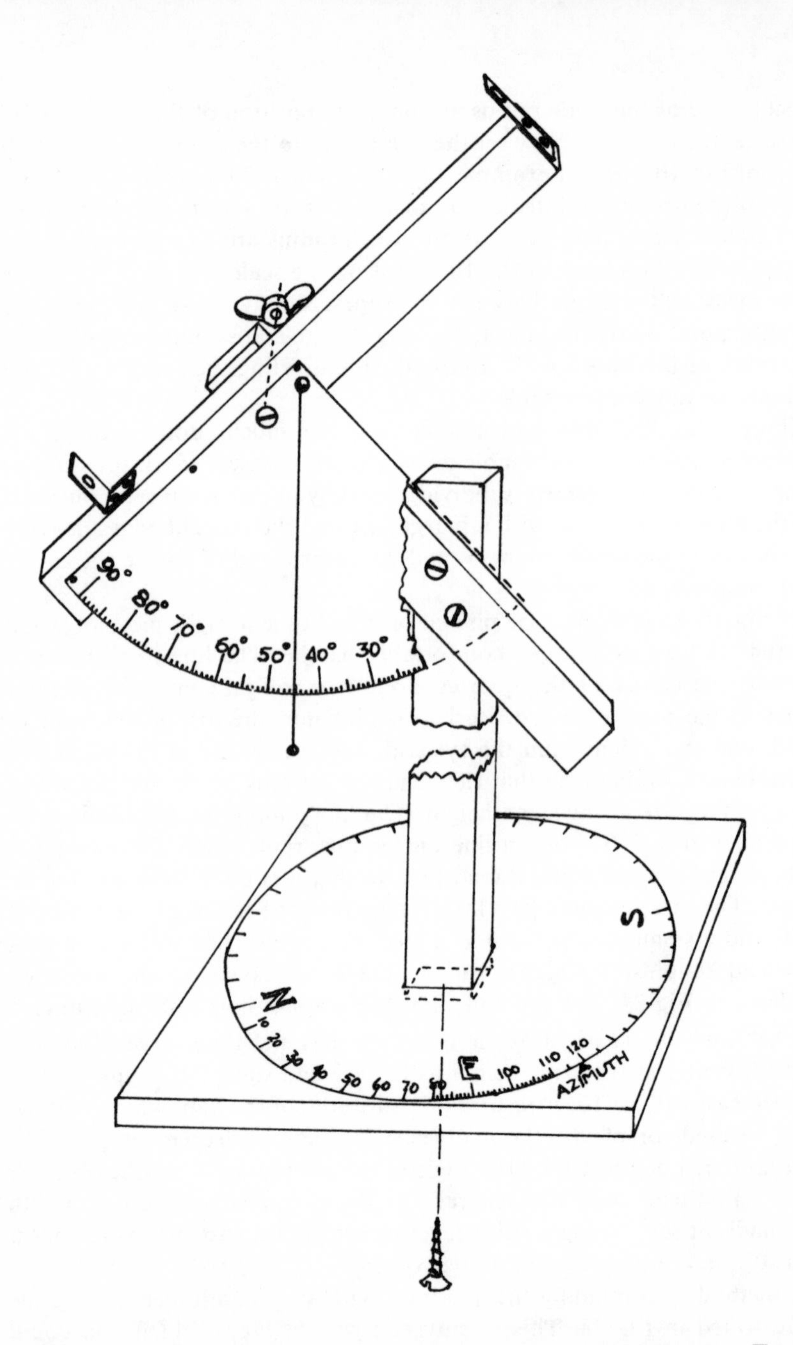

Fig 38 A combined quadrant and azimuth measuring instrument. To allow readings near the zenith, the support for the quadrant itself must be offset to allow the eye to be placed behind the lower sight.

of the quadrant, and yet hangs alongside the scale. When hand held, you point the sights at the object in question, and firmly press the plumb line against the scale when the plumb line is hanging steadily. You can then take the instrument from the eye and note the reading. With the stand, this procedure is much simplified, and likely to be more accurate.

To make the arrangement shown in Fig 38, take a piece of wood about 2cm thick and 23cm square, to act as the base. The quadrant itself is supported on a piece of wood about 3cm square and of length sufficient to hold the quadrant at a suitable height for use, say 40cm. A short length of this upright support is screwed at an angle at the top so that the quadrant is offset from the centre of the base, allowing the eye to be placed behind the sights when the instrument is pointing directly upwards. The quadrant itself is secured to this offset arm of the support by a countersunk bolt passing horizontally through the arm of the quadrant and through the support, and fastened by a wing nut on the back. This allows the instrument to be inclined as required for different altitudes. The bottom of the vertical support piece must be cut squarely and is secured to the base by a single wood screw from underneath, allowing it to be turned in azimuth.

An azimuth scale, as large as the base allows, must be made on thick card, and marked from 0° to 90°, 180°, 270°, and back to 0° in 1 degree intervals anticlockwise, numbering each 10 degree interval. Cut a square hole in the centre of the scale so that it will fit over the vertical support. Being a square hole, the scale will then turn with the support. It would be a good idea to glue the scale card to the support to prevent any relative movement. Now make an index mark opposite 0° on the azimuth scale.

To use the mounted instrument, set 0° against the index on the azimuth scale, and tilt the quadrant to set the altitude reading on the altitude scale to your latitude. Without disturbing these settings, line up the sights on Polaris. If you wish to be as precise as possible you can make the necessary correction for the true position of the pole as described in Chapter 1. The instrument is then ready for use. Without moving the base, turn and elevate the quadrant to line up the object in the sights. Note the time and read both scales (it is most important to note the time immediately). You can now see why we marked the azimuth scale anticlockwise in the opposite sense to the compass card so that east is to the left of north: it is because the whole scale is meant to rotate instead of a pointer moving above a fixed scale card. If you have devised an azimuth indicator with a moving pointer over a fixed scale on the base, you must calibrate the scale clockwise.

USING THE QUADRANT

Of the following stars at least one should be visible from the northern hemisphere this evening. (At what times of the year, roughly will each be prominent in the sky?) Sirius; Regulus; Arcturus; Altair; Algol. Measure the altitude and azimuth of at least one of these at 21h. What is the sidereal time? What are the RA and Dec of the star?

Here is another exciting exercise. From a suitable ephemeris (see Reading List, p 201) find out when Venus is at a reasonably wide elongation from the Sun. You will need Venus's RA and Dec for the day on which you are going to carry out the experiment, either using the methods we will study in a later chapter, or from a handbook. With the globe, or the astrolabe (see the end of this chapter) calculate its altitude and azimuth at noon. Yes 12h 00m! Set up the quadrant and azimuth instrument well in advance of noon. You could align the azimuth scale by the Sun, from a measurement on the globe of its azimuth at a particular time during the morning, or you could set it by Polaris the night before. At noon with the instrument set to Venus's altitude and azimuth, look carefully along the sights of your quadrant. With any luck, there in the middle of the day you will see Venus. You may have made some errors, so look around the immediate vicinity of the sky where the instrument is pointing.

The Astrolabe

The astrolabe was used for navigation until relatively recent times, and its invention is generally credited to the famous ancient Greek astronomer Hipparchus who lived a century and a half BC. It is somewhat more complicated than the other instruments we have made in this chapter, so perhaps this will help increase your respect for the astronomers of over 2,000 years ago!

The astrolabe fulfils many of the functions of a celestial globe, in some cases with greater accuracy, in others with less. The main limitation of the astrolabe is that, strictly speaking, it is suitable only for the latitude for which it is made. In practice, accuracy is not unduly affected for a reasonable range of latitudes. On the other hand, the materials you will need are much easier to come by, and the precision with which you can mark the instrument is much greater than a globe.

What we are going to do is to project both the celestial and the local co-ordinates (RA and Dec, altitudes and azimuths), on to the same plane. The particular method of projection is called stereographic projection. How would the celestial sphere appear as seen from a pole P in Fig 39, which is immediately beneath the centre of the equator? If we imagine the ecliptic on the sphere, it will be half above the equator, with its highest point at *d*, and half below with its lowest point at *a*. To project the ecliptic on to the plane of the celestial equator, the plane of projection, we can

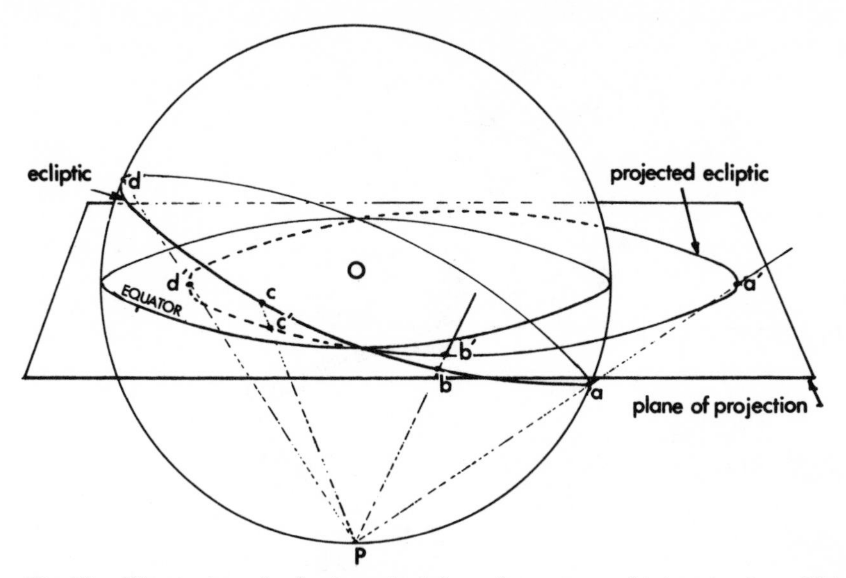

Fig 39 Illustrating the basic principles of stereographic projection. This shows how the ecliptic is projected from the celestial sphere on to the plane of the equator.

see from Fig 39 that the point *a* will appear to be at *a'*. The point *d* will appear to be at *d'*, and some of the intermediate points, *c* and *b*, will be at *c'* and *b'* respectively. The celestial equator itself is on the plane of projection so that it will be a circle of exactly the same size on the projection, but the ecliptic will be a circle larger than the equator and with a different centre. Obviously we can't work with three dimensional models such as Fig 39 represents, so we project the ecliptic as shown in Fig 40.

The circle at the top is a view along the equator, the plane of projection, so that the equator is a line and the celestial sphere is drawn as a circle. Immediately beneath the centre of the upper circle at O we draw another circle of the same diameter with the centre at O″ which is vertically beneath O. We then draw a horizontal line through O″ which is parallel to the equator in the upper circle. Now draw the ecliptic through O in the top circle at an angle of $23\frac{1}{2}°$ to the equator so that it cuts the circumference of the circle at *a* and *d*. Drawing faint construction lines (because we will erase them later) from P to *d* and through *a* to the equator gives the points *d'* and *a'* which are on a diameter of the projected ecliptic. We can then drop verticals from *d'* and *a'* till they meet the line through O″, giving *d″* and *a″*. These two points lie on the ecliptic, which will be a circle on our diagram. To find the centre of this circle, we measure the

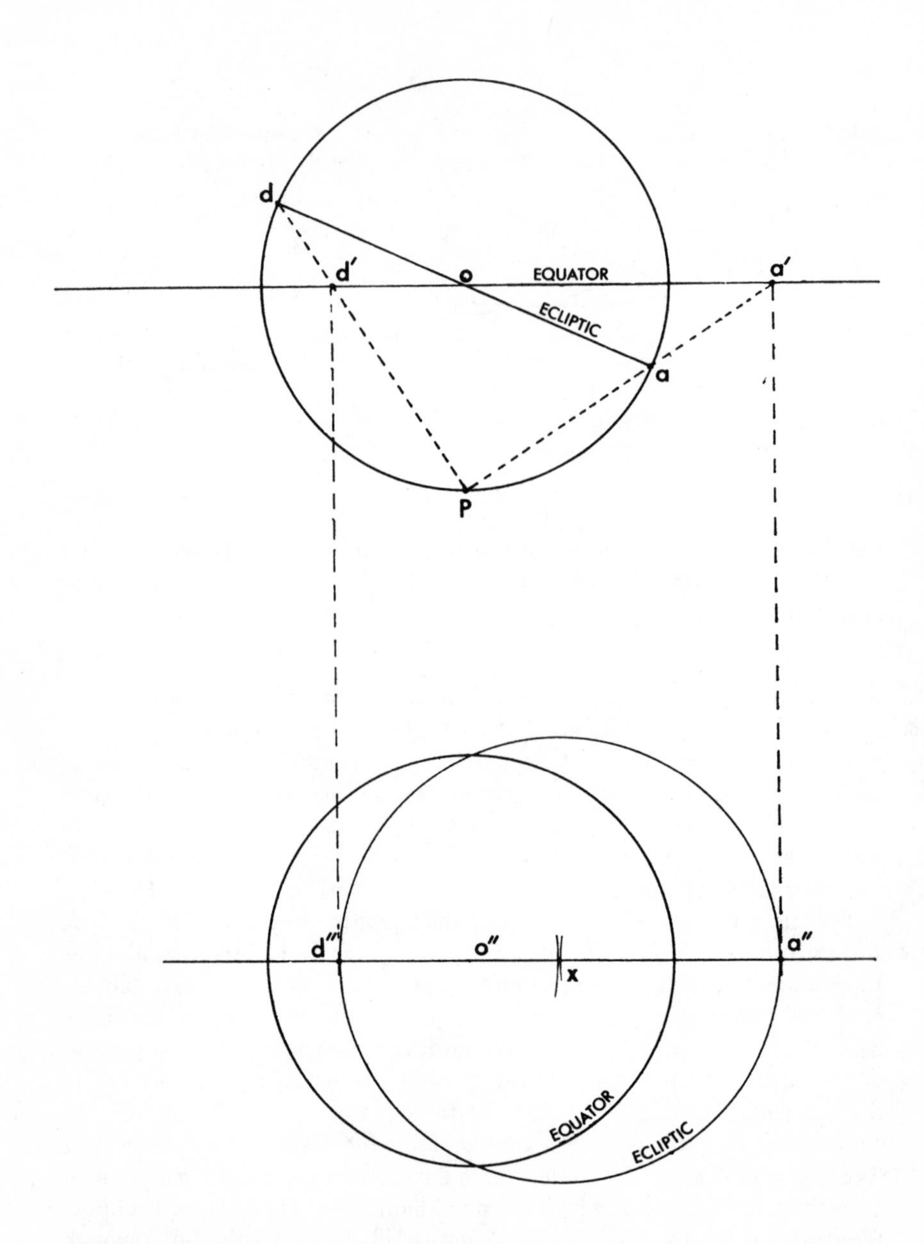

Fig 40 The same projection as Fig 39, showing the method of drawing the projection.

distance from a'' to d'' and, since this is the diameter of the projected ecliptic, divide by two to get the radius. We then set the compasses at this radius and strike two faint arcs, one from a'' and the other from d'', as shown in the diagram. We can then judge precisely the position of the centre x, half way between the two arcs, checking it with the compasses to make absolutely sure before drawing the ecliptic circle on our projection.

An alternative way of making the projection from the top circle to the lower circle is to measure the distances to both a' and d' from O in the upper diagram with the compasses and transfer the measurement to the projection by striking off the arcs for each distance from O″. This is less wasteful of paper, but much more prone to errors in the translation of the measurements, and until you are familiar with the process it is better to use the vertical projection method.

The astrolabe is made of two main parts. The base, called the *tablet*, is to be marked with the horizon, azimuths, and the altitudes. The second part, called the *rete*, is a transparent disc which rotates about the pole on top of the tablet, and is marked with the ecliptic, the hour circles and the circles of declination. On the rete we can also mark the principal stars, since these are fixed relative to the lines of RA and Dec.

The original astrolabes had retes made from an intricate lattice of brass, many of which were decorated with elaborate patterns. We can use plastic sheet, such as perspex (which is fairly expensive) or a more flexible plastic such as acetate, sold in artists' materials shops. It is a good idea to make the drawing for the rete on a piece of paper first, then place the plastic sheet on top, fasten it in position temporarily, and mark the plastic with a fine sharp point. A pair of dividers or, better still, a set of basic draftsman's instruments is needed to scribe the circles.

The method of making the stereographic projection of these celestial circles and lines is illustrated in Fig 41. You must first decide the maximum diameter you can afford for the declination which falls on the horizon. The diagrams in this chapter are drawn for latitude 52°N, so the most southerly declination we need is 38°. Taking 40°S as the nearest round figure, the outside diameter of the instrument will correspond to Dec −40° and, as a rough indication, the corresponding diameter of the equator is just under half the outside diameter at this latitude.

Draw a circle to represent the side view of the celestial sphere as before, with an appropriate diameter, and draw in the equator and ecliptic. From the centre, measure a series of angles above and below the equator for every circle of Dec required. The most southern declinations will be well spread out, so it is advisable to mark angles at 10° intervals at least below Dec +40°. You can then draw the 'side view' of the circles of declination, parallel to the equator. For the purposes of the construction you need only

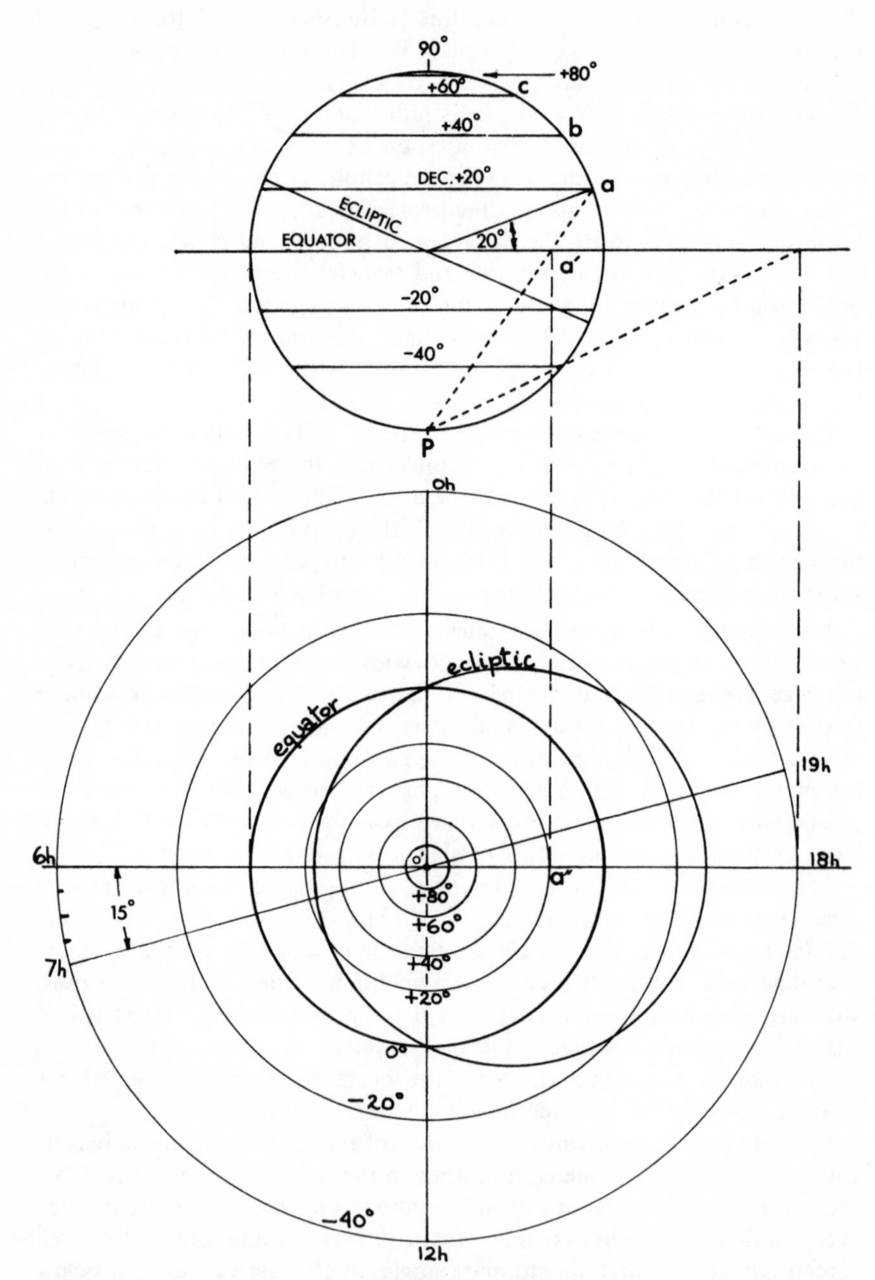

Fig 41 Stereographic projection of the circles of Dec and RA from the celestial sphere on to the plane of the equator.

mark the points at which the constructional lines at angles of 10°, 20°, etc to the equator cut the circumference (ie points *a*, *b* and *c* in Fig 41) because it is these points which are to be projected on the lower circle. These points mark the radius of the declination circles, the centres of which are all at the pole. Since the pole is immediately above the point of projection it must also be in the centre of the projection. To avoid possible confusion between the ends of the ecliptic and the declination circles it is a good idea to draw the ecliptic in full on the top diagram.

Project each end of the declination circles, such as point *a* for the 20° circle, to obtain *a″* on the rete. Drop the perpendicular from *a′* where the line from P to *a* crosses the equator, to meet the 18h line at *a″* on the rete. Set your compasses to radius $O-a''$, and draw the 20° circle with the centre at O. Repeat this procedure for the other Dec circles, then project the ecliptic as already described. The hour circles can then be drawn as straight lines from O with the hours of RA marked at their ends as shown. The line for each hour is 15° from the adjacent hour circle line. Note that the RA of the ecliptic at its highest point is 6h 00m, and that the hours are marked anticlockwise (clockwise for a southern sky astrolabe). Mark in the brightest stars at their appropriate RA and Dec (see appendix B).

The markings can now be traced on to the plastic sheet by scratching firmly with a sharp point, and the marks filled in with indian ink or ball-point pen to make them easily visible. For greater accuracy in reading the hour circles, each hour can be divided into 5, 10, or 15 minute intervals around the circumference. When you have marked out the rete you should notice that the stars are in a mirror image of their usual relative positions, as happened with the globe. Once again, we are looking at the sky from the outside.

Now we come to the tablet. This can be marked out directly on to paper which will later be glued on to a board, or it can be marked on stiff artist's quality card. You will need an area at least as big as the outer circle of the rete, and you must take care in positioning the diagram to ensure that as much as possible of the constructional work can be carried out on the piece of paper. Start by drawing a circle exactly the same diameter as the celestial sphere on the rete. A horizontal line through the centre in this case (see Fig 42) is the equator, and the circle itself represents the meridian. This time, we want to mark the circles which show points above the horizon at the same altitude. These circles are called *almucantars,* and each almucantar is parallel to the horizon. Since a vertical line through P and the centre of the circle is at right angles to the equator, it must pass through the pole. We can draw in the horizon for our latitude using the fact that the altitude of the pole is the same as the latitude. We draw the horizon line through the centre of the circle making an angle equal to our latitude

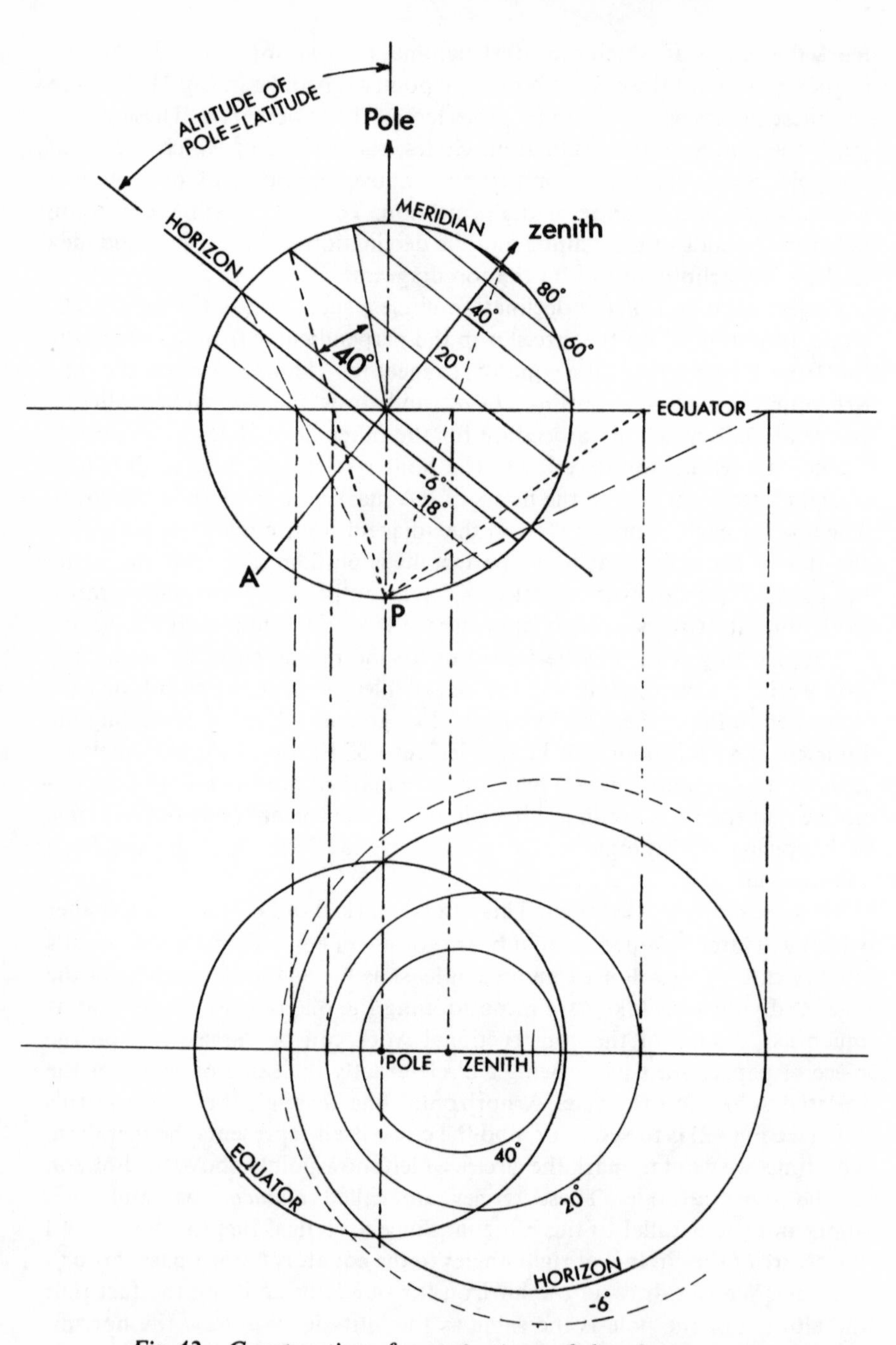

Fig 42 Constructions for projections of the almucantars.

with the line through the pole (nb not with the equator).

Draw a line from the centre of the circle at 10° *from the horizon,* and another at all the other altitudes for which we wish to draw almucantars on our tablet. At each point of intersection with the circle we draw a line parallel to the horizon. We draw similar lines for altitudes of 6°, 12°, and 18° *below* the horizon. These 'negative' altitudes will be used to indicate when the Sun is below the horizon by these amounts, corresponding to the ends of civil, nautical, and astronomical twilight respectively.

Fig 42 shows, at the top of the diagram, a meridian with the horizon drawn for a latitude of 52°. The construction of the 40° almucantar is shown in some detail, with the basic construction line at 40° to the horizon, and the corresponding parallel. Parallels for some of the other altitudes have been drawn in without this construction shown. The zenith is drawn at right angles to the horizon, through the centre of the circle. The projections are then made of the zenith, and both ends of the lines indicating the horizon and each altitude.

For the horizon and each of the almucantars we shall obtain two points on the meridian. For each of these we must estimate the centre with compasses as we did for the projection of the ecliptic, and the construction arcs are shown for the horizon. Having found the centre, we can draw the circle and this procedure is followed for each almucantar. In the case of the altitudes below the horizon for twilight duration times, we need not draw the whole circles, since the Sun cannot rise or set further south than south east or south west from this latitude. Note that the finished almucantars are not concentric. This stage is completed by projection of the zenith, although this is only a check for the next stage.

The projection of the azimuths must be carried out on the same diagram as the almucantars, but, since this is a fairly complicated stage, the method is shown alone in Fig 43. The finished tablet will have both azimuths and almucantars as illustrated in the lower diagram in Fig 44. Returning to Fig 43, we can see that the immediate projection of an azimuth such as the one drawn to the right of the zenith is impossible, since this is only a perspective view on a flat page of a line which should be drawn in three dimensions. To project the azimuths we must turn to another view of the sphere, the view down on to the zenith. Just as each line of longitude at Earth's poles makes an angle with its neighbours equal to the difference in longitude, so each circle of azimuth crosses the others at the zenith at an angle equal to the difference in azimuth. There are two azimuths on our original diagram which we can use to find the others. These are the north south meridian (azimuth 0° and 180°) and the east west azimuths (90° and 270°). The east west azimuths are represented in the top circle of Fig 43 by the line joining the zenith through the centre to the point *x*, at

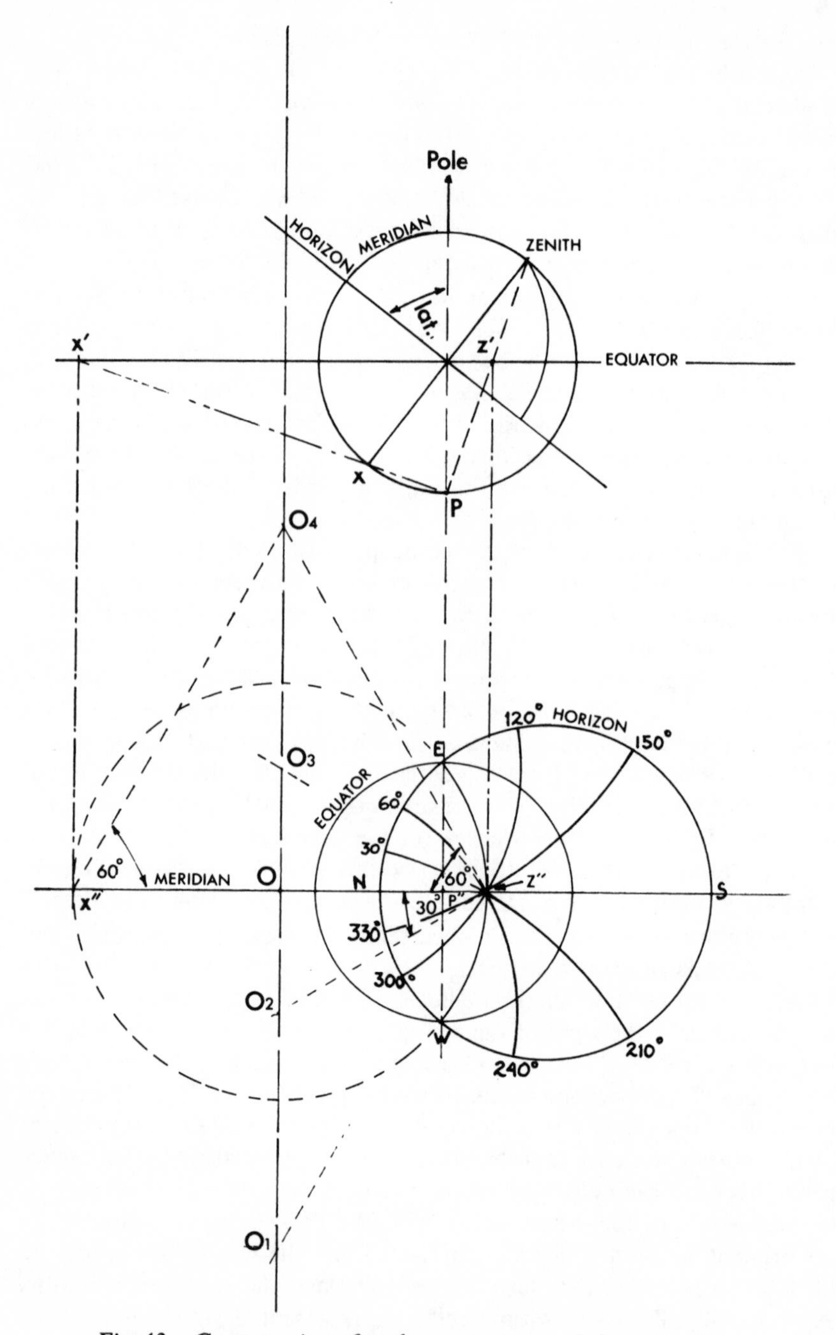

Fig 43 Constructions for the projections of the azimuths.

an 'altitude' of −90°, directly beneath our feet.

This circle can be projected on to our tablet in the usual way, as indicated in the drawing. The portion of this circle which comes inside the horizon (and which should pass exactly through the two points where the circles of the equator and horizon intersect, and through the zenith) are the azimuths 90° and 270°—the east and west azimuths. Now all we have to do is to construct circles which cross this arc at the zenith at the required angles.

Take for example the azimuth 150° and 330°. This makes an angle with the east–west azimuths of 150° − 90° or 330° − 270° = 60°. Now a tangent to each of these circles at the point of intersection at the zenith will make an angle of 60° and the lines at right angles to these tangents must pass through the centres of their respective circles.

Since the circles pass through the zenith, we can draw a number of lines at angles to the meridian of 10°, 20°, 30°, and so on up to 80°, from the projected zenith, Z'', and from the opposite point x'', since the same azimuths must make the same angles where they cross under our feet as overhead. Where these two lines intersect will be the centre of the projected azimuth circle and the distance from such a point, O_4 in the diagram, to the zenith will be the radius. Any pairs of lines drawn from x'' and Z'' at the same angles to the meridian must intersect on a line which passes through the centre, O, of the 90°–180° (east–west) azimuth circle projection, and which is at right angles to the meridian.

To draw the remaining azimuths, therefore, we can construct a line through O at right angles to the meridian and measure angles from Z'', and hence locate the centres O_1, O_2, O_3, etc. From each of these we draw an arc of a circle of a suitable radius to pass through the zenith from one side of the horizon circle to the other.

You will come across a practical difficulty when you draw the azimuths at above about 60° to the east–west circle. You can see from Fig 43 that the 60° line intersects at O_4, some distance from O on the meridian. The 70° line is quite a bit further still, and the 80°—well! Your compasses will not stretch that far. There is nothing for it but temporarily to secure the tablet to a large table and find the centres as accurately as you can with a very long straight edge, then, with the help of an accomplice, draw your circle with a pencil tied near the point to a piece of string. Your assistant holds the other end of the string at the centre you have just found and, holding the string taut, carefully draw the arc required.

Having marked the azimuths in degrees from 0° at the north, through 90° (east), 180° (south), 270° (west), and so on back to north again and similarly having marked the almucantars from 0° at the horizon to 90° at the zenith, the astrolabe is at last ready for the first stage of assembly.

Place the plastic rete on top of the tablet so that the poles on both are exactly together. If the equator is marked on both, these circles will also coincide exactly. Any star appearing inside the horizon circle would be visible for the time corresponding to the particular sidereal time to which the rete is set. This is the right ascension (read from the scale on the rete) on the meridian (azimuth 180°, or due south) on the tablet.

We will add a scale showing every day of the year round the outer circumference of the tablet which can be used to indicate UT (or mean time) in conjunction with the hour circles on the rete. To do this it is best to first glue the tablet on to a thin board, if it has not been drawn on a thick card in the first place. You must now temporarily insert the pivot for the rete to help in marking the date scale. Carefully press a drawing pin squarely through the pole on both the rete and the tablet, into a scrap of wood behind the tablet if it is on card. Now the rete can be turned about the pole to represent the diurnal rotation of the Earth.

The hour circle on the rete which lies on the meridian on the tablet gives the sidereal time. So if we turn the rete in suitable steps, making a mark on the tablet at the 00h mark on the rete at each step, we can put a date against each mark corresponding to the date at which the sidereal time on the meridian is 00h UT. For example, set 00h on the meridian; now make a mark opposite 00h on the circumference of the tablet (also on the meridian, of course). This is September 23, when 00h sidereal time corresponds to 00h UT. Now turn the rete to read 00h 17m on the meridian. The mark against 00h now corresponds to September 26. Carry on in the same way every 4 days, using the sidereal time table in appendix B. When we have an RA of 06h 00m on the meridian, the date opposite 00h is December 22, and when we have 12h sidereal time on the meridian, 00h UT is on March 23, and so on. Finally divide the 4 day intervals into 4 to give a point for each day of the year. Only the positions for the first day of each month, and the solstices and equinoxes, are shown in Fig 44, which illustrates the finished rete and tablet with some of the lines and circles omitted for extra clarity. To read local sidereal time (ie the RA on *your* meridian) set the mean time corrected for your longitude east or west of the centre of your time zone against the date on the tablet. Unfortunately there is nothing you can do about very different latitudes, and you will need a new tablet for these.

One last touch to complete the astrolabe. Cut a thin strip of the plastic you used for the rete, just a little longer than the radius of the rete. Scribe a fine line along its centre, mark intervals of declination along the line at the same spacings as on the rete, and fill it with indian ink or ball-point pen. Push the drawing pin through this line about 5mm from one end and then reassemble the astrolabe. The free end of this pointer will just reach

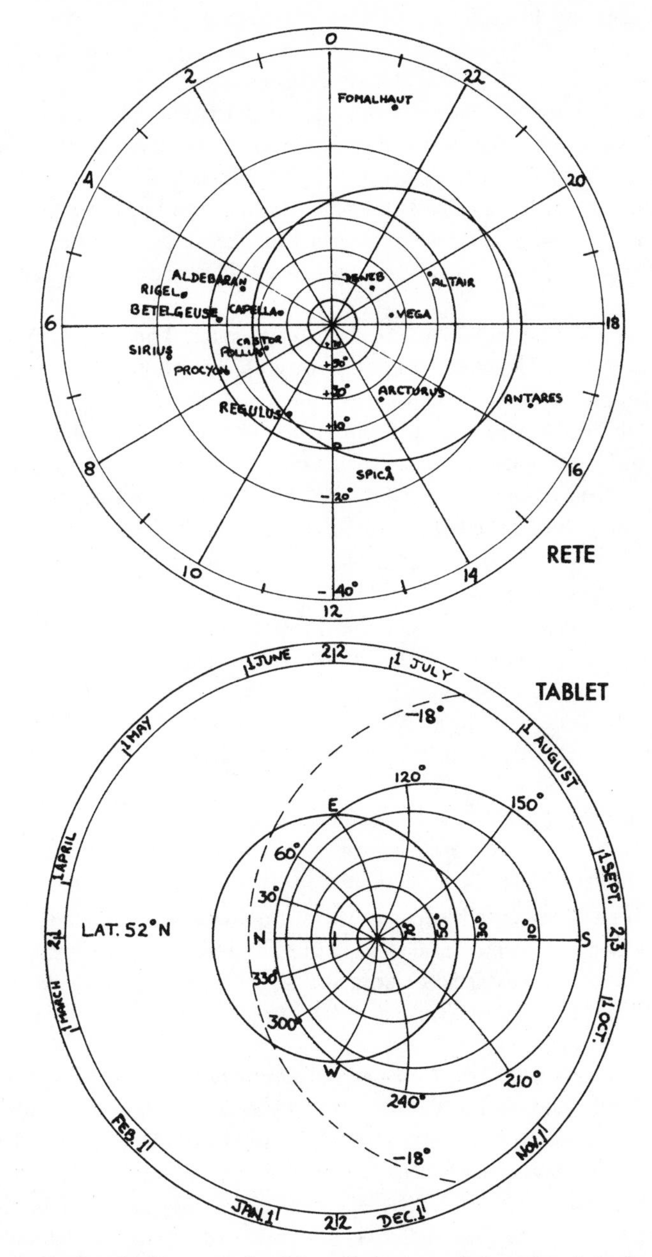

Fig 44 The finished rete and tablet of the astrolabe, showing the principle lines of calibration on both. Note carefully the direction of the hours of RA around the rete, and dates around the tablet.

the outer scale of dates.

USING THE ASTROLABE

Here are some exercises worked out on the astrolabe, and there are some which you should do for yourself at the end of the chapter as usual.

1 A close approach of the Moon to both Mars and Venus (in northern Europe the Moon occulted Mars) was predicted in the almanac for 1972 May 15 at 20h 44m. Did this interesting conjunction take place in a favourable position in the sky from Greenwich? The RA of Mars was 6h 7m.

First we must find out whether the Sun has set. The correction for the equation of time on May 15 is small enough to ignore for this example, so we set 12h on the rete against the date, and place the pointer on the meridian where the Sun is at 12h 00m UT. The RA of the Sun is 3h 28m, and this is also the sidereal time. The Dec of the Sun is where the line on the pointer crosses the ecliptic (at about +19°). Now turn the rete and pointer together until the time against the date is the predicted time of the conjunction, which is at 20h 44m UT. Using the twilight almucantars we find that the Sun is below the civil twilight line but not yet on the nautical twilight line. It is about 9° below the horizon. That's not too bad: the sky will be fairly dark away from the Sun's azimuth, which is about 312°. But where is Mars? With the rete in the 20h 44m position, move the pointer to the RA of Mars, 6h 7m and, assuming Mars is on the ecliptic, we find it is at altitude 18° azimuth 285°, which should be far enough away from the Sun.

2 When does Rigel set on February 3 at Greenwich?

Set the rete so that Rigel is on the western horizon, and read the time against February 3; 1h 34m. The sidereal time is 10h 25m.

3 On June 8 we see Arcturus is about due west at an altitude of 30°, as near as we can tell. What time is it? Turn the rete until Arcturus is at this altitude above the western horizon. At 30° exactly it is only 6° south of west, and the time against June 8 is 1h 30m, so this was the approximate time.

4 Use the astrolabe together with the quadrant to find Venus in broad daylight. First, find when Venus is reasonably far from the Sun, using either an ephemeris, or the methods we shall study in chapter 8. In 1975, for example, the planet was a brilliant evening star in the spring and early summer. We will try to find it before sunset on May 28; we find that the RA is 7h 28m and the Dec is $+24\frac{1}{2}$°. Let's make it really spectacular and look for the planet at noon as suggested earlier. Where will Venus be then?

Set 12h 00m opposite the date on the tablet. Now set the pointer to

Venus' RA, 7h 28m, and at a point corresponding to declination 24½° read the altitude and azimuth; we find it is altitude 45½°, azimuth 109°. So these are the local co-ordinates for latitude 51° at noon local mean time. Now we set up the quadrant, suitably oriented with its azimuth scale (using the Sun to set it up if necessary, but don't forget the equation of time) and we incline the instrument to altitude 45½°. Now look carefully along the sights and, if all is well, there you will see the bright spark of Venus against the blue of the daylight sky. In case you have made any small errors, it may be necessary to search a small area of sky centred on where the quadrant sights are pointing, but you should not be more than a degree out.

5 Orion's Belt when rising and setting indicates east and west respectively. How do we know at a glance which is which, from moderate latitudes? Turn the rete until the belt stars are on the eastern horizon. You will see that delta Orionis is almost exactly at azimuth 90° as it rises, and by the time the whole of the Belt has cleared the horizon, the azimuth is about 95°. But notice that the Belt is from 265°–270°, the three stars are parallel to the horizon.

These are only a few of the demonstrations and practical problems that can be worked out on the astrolabe. Most of the problems in this book concerned with position can be checked on the astrolabe, or the globe, and they form useful rapid reference devices to check the answers to more precise calculations.

For those who want their astrolabe to resemble the ancient instruments more closely, you can add a diametrical arm, pivoted in the centre, on the reverse side. This is called an *alidade,* and should be fitted with sights at each end. Then you simply mark an altitude scale on the reverse side, and place a loop of string at the altitude 90° point. This is used to suspend the instrument from one hand while you take an altitude reading with the alidade. You will not get such accurate results as with the quadrant, but it will enable you to measure altitudes approximately and plot the position of the object on the rete and tablet.

EXERCISES ON CHAPTER 6

1 In a group of 3 stars, how many observations must you make with the cross staff to be able to draw the group to scale?

2 Take an altitude reading of a star with the quadrant, then try to measure its angle above the horizon with the cross staff. Which is likely to be the most accurate reading and why?

3 On March 18 you spot a spectacular fireball in the sky which disappears at an altitude of about 40° in the south east. What is the RA and Dec of this point,

approximately, if (a) the time was 21h 00m UT and (b) if it was ten minutes past midnight? (The answer in the appendix is given for astrolabes made for approximate latitude 51°N.)

4 With the quadrant measure the altitude and azimuth of the Sun on the same date (the next year if necessary) as you carried out the stick shadow experiment in Chapter 1. How closely do the results compare? Now predict the same set of measurements on the astrolabe.

7

Measurement on the Celestial Sphere

When we measured the angular motion of the Sun across the sky in Chapter 1 by plotting the movement of a shadow, we found that it was the same each hour. We now know that the Earth is rotating on its axis once in 24 sidereal hours so that the Sun and other celestial bodies appear to turn about the celestial pole at the rate of 15° per hour. But the angle measured by the shadow movement was less than 15° per hour, unless the experiment happened to be carried out close to one of the equinoxes when the Sun is on the equator.

The reason for this can be found from Fig 45. On the left is a side view of the celestial sphere, with the observer at the centre, O. A star S is shown

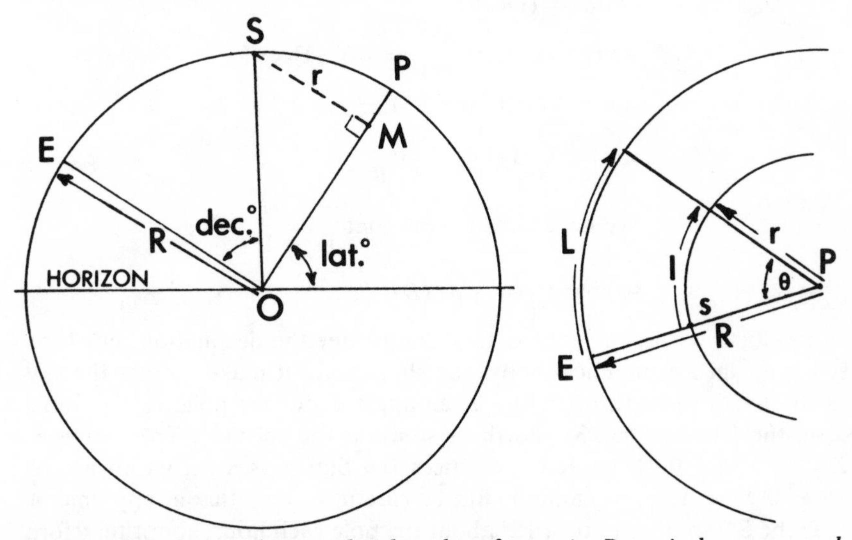

Fig 45 Calculating the angular lengths of arcs in Dec circles measured between hour circles of RA, at different declinations.

at an angle to the equator, which is its declination. If we could look at the sphere from above the pole, as shown in the right hand diagram, we would see that while the whole celestial sphere turns through an angle O, a star on the equator starting at point E will pass along an arc of length L, while the star S passes along a shorter arc of length *l*. The two arcs, lengths L and *l*, subtend different angles to the observer at O, so they are not the same as the angles measured round the celestial pole.

The arcs are parts of the circumferences of two declination circles of radius R and *r*, as shown in the diagram. Since the length of an arc is the same proportion of the circumference as the angle it subtends at the centre of the circle is to the angle right round the centre, that is 360°, we can say that;

$$l = 2\,\pi\,r \times \frac{\theta}{360}, \text{ and } L = 2\,\pi\,R \times \frac{\theta}{360}.$$

Dividing one equation by the other we find that

$$\frac{l}{L} = \frac{r}{R}$$

Referring to the left hand diagram in Fig 45, we can see that since SM is parallel to EO, the angle OSM is equal to the angle SOE, which is the star's declination.

Since

$$\text{OSM} = (\text{Dec})^\circ,$$

we can say that $\frac{SM}{SO} = \text{Cos (Dec)}^\circ$.

But SO = R, and SM = r,

so that $\text{Cos (Dec)}^\circ = \frac{r}{R}$.

We have already seen that $\frac{r}{R} = \frac{l}{L}$,

so that $l = L\ Cos\ (Dec)^\circ$.

In practical terms, this means that the greater the declination, north or south, of an astronomical body, the shorter arc it makes across the sky as the Earth turns through any given angle about the pole. In the Sun's case, the Dec can be 23½° north or south at the solstices. The cosine of 23½° is about 0·92, so at the solstices the Sun passes across an arc of 15 × 0·92 = 13·8° in an hour. But be careful to note that at any time of year the Sun must still turn 15° about the pole each hour, and it therefore turns at 15° per hour around the style of our sundial, because this points at the pole. The further the stars are from the pole, the longer is the arc they make. This distance from the pole is another angle, and is clearly the complement of Dec (90 −Dec)°. This distance is called the *co-declination*, or *polar distance*. Some of the other co-ordinates with which we have now become familiar, such as latitude and altitude, have complementary co-ordinates and often these are more useful than the ones we have used up to now.

In Fig 46, for example, a three dimensional view of the celestial sphere is shown. A star S is shown high to the south east of the observer at O, which is the centre of the sphere. The star's altitude is the angle above the

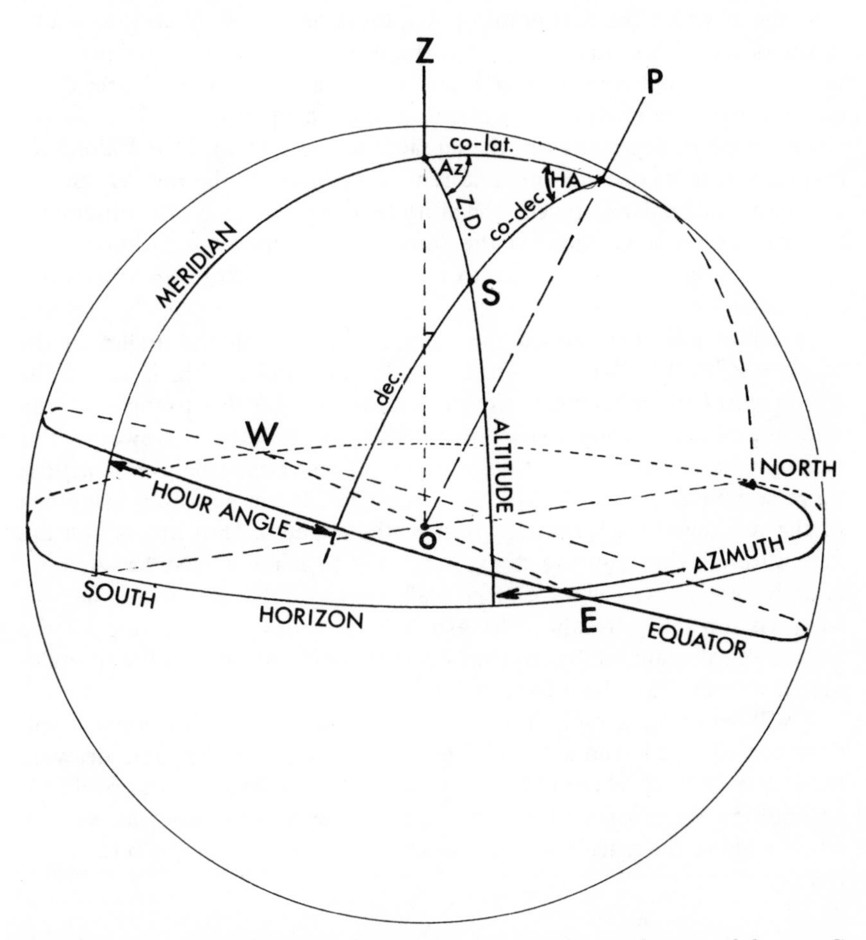

Fig 46 The celestial sphere, showing the principal co-ordinates of the star S, and the astronomical triangle ZPS. Note that the hour angle, HA, is marked to the east of the meridian in this diagram, hence is negative.

horizon, measured along a vertical circle which passes exactly overhead (through the zenith) and through the star itself. The distance along this circle from the star to the zenith is the complement of altitude, the *zenith distance*. The angle this circle makes with the observer's meridian is the star's azimuth. This can be measured either as the angle (from the north) the circle and the meridian make as they cross at the zenith, or the arc from the north point of the horizon to the point where the line from the zenith through the star meets the horizon.

The hour circle of the star passes from the pole through the star, and meets the celestial equator at right angles. Unless we know the time, and the time at which the first point of Aries was on the meridian (due south of the observer) we cannot say what the star's RA is, but we can measure the hour angle between the star's hour circle and the observer's meridian, either where they cross at the pole, or along the equator.

The angle in degrees, from the equator to the star, measured along its hour circle, is its declination. The remaining angle to the pole is the co-declination, or polar distance. The altitude of the pole above the observer's horizon, as we have seen, is the same as the observer's latitude. The remaining angle, from the pole to the zenith, is therefore the observer's *co-latitude*.

We have now determined the sides and nearly all the angles in the triangle ZSP. This triangle, from star to pole and zenith, is called the *astronomical triangle*, and if we know how to solve this triangle we can find one of the unknown sides or angles from the others. However, this is a spherical triangle and the geometry of spherical triangles is a little more complicated than that of plane triangles. It can be solved using one of the mathematical formulae given in the appendix, but unless you are very sure of what you are doing it is easy to make a mistake. A more straightforward way is by a large scale stereographic projection such as we used for the astrolabe. The astrolabe is a device for giving all the astronomical triangles for any particular latitude. By tackling the triangles alone, we can draw them for any latitude.

It will help considerably in understanding the rest of this chapter if you have already made the astrolabe, because we will use the same drawing techniques, but we will omit the constructions not relevant to our problem. In addition we must measure the scale of the drawing and, as we saw when making the astrolabe, this varies considerably with the position in the sky.

Scales on a stereographic projection

The astrolabe was a stereographic projection made by projecting lines from a diagram in which the pole was the centre. This produced a diagram in which the lines of RA were straight, and radiated from the pole, but the lines of azimuth were arcs of circles. The declinations were shown by circles concentric about the pole, but at a non-uniform distance from it. The circles of altitude were neither uniform in radius, nor concentric.

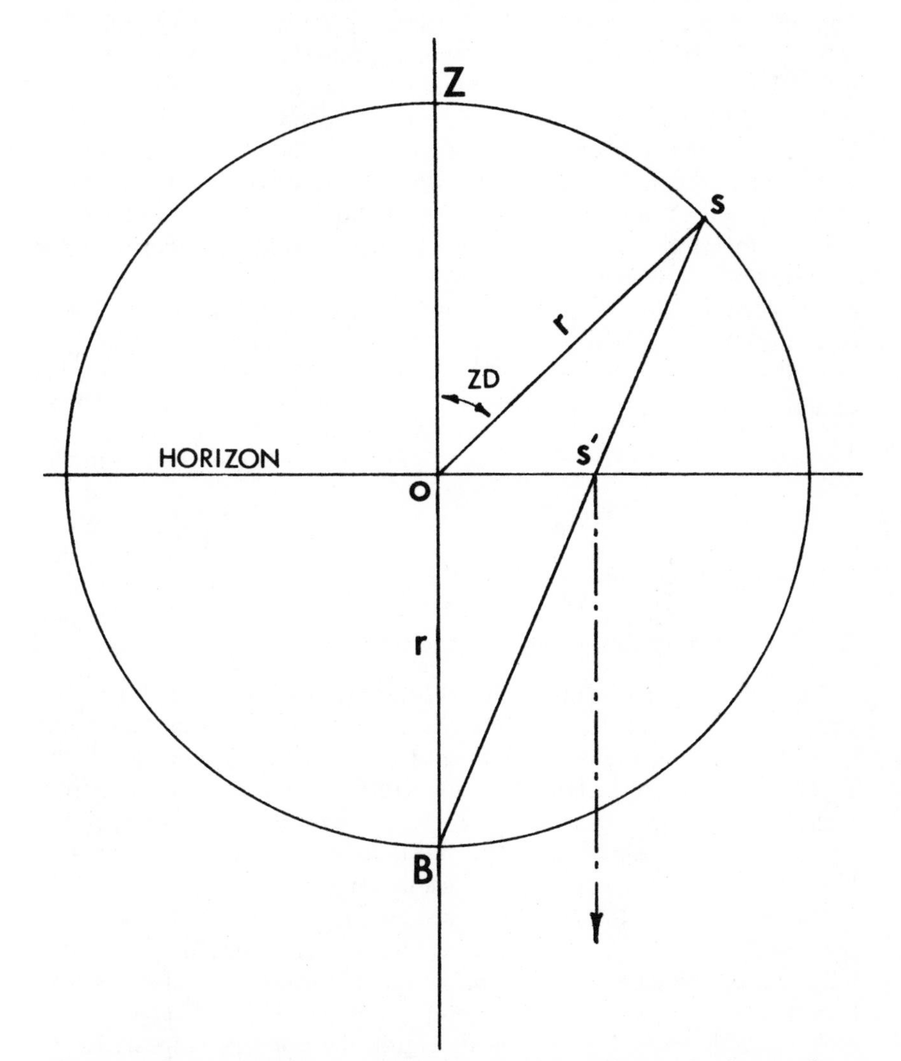

Fig 47 The celestial sphere, centred at the observer's zenith, showing how a stereographic projection would be made for the zenith distance circle of the star S, and how the length of the projected radius OS′ can be calculated.

If the original diagram from which the finished projection had been made had had the zenith in the centre, the non-uniformities would have applied to the celestial co-ordinates, while the observer's local co-ordinates would have been straight lines and concentric circles.

We will work out the scale using a zenith centred projection. We start with a side view of the celestial sphere as shown in Fig 47. We want to find the zenith distance of the star S, when projected on our diagram. This is the length OS′. Note that it is measured along the meridian in the plane of the horizon. All our measurements with stereographic scales must be made along the meridian. We would have dropped a perpendicular from S′ to our projected diagram and drawn a circle about Z of radius OS′, if we were making the whole drawing. But we can see that in triangle OSB, OS = OB = r, the radius of the projected horizon. The angles OBS and OSB are therefore equal, and we will call them b. Adding the angles in the triangle OSS′, we have SOS′ + OSS′ + SS′O = 180°. Now, OSS′ = b, and SOS′ is the altitude of S, which is 90° – zenith distance (ZD). Therefore, angle SS′O = 180 – (b + (90 – ZD)). But angle SS′O is also equal to (180 – OS′B) so that angle OS′B = 90 – ZD + b. We know all the remaining angles in triangle OS′B, so we can add them and equate to 180°.

This gives: $90 - \mathrm{ZD} + b + b + 90 = 180°$.

Therefore $2b = \mathrm{ZD}$, or $b = \dfrac{\mathrm{ZD}}{2}$.

But, in triangle OS′B, $\dfrac{\mathrm{OS'}}{r} = \mathrm{Tan}\ b$

$\mathrm{OS'} = r\ \mathrm{Tan}\ b$, so that finally we have $\mathrm{OS'} = r\ \mathrm{Tan}\ \dfrac{\mathrm{ZD}}{2}$.

This means that to draw a line on our stereographic projection to a scale which relates the length of the line to the angle it represents, measured from the centre of the projection (in this case, the zenith), we divide the angle by 2, look up the tangent of the half angle in some trigonometrical tables, and multiply this figure by r, the radius of the projection. The simplest radius to use is, of course, 10 units, for example 10cm or 10in. We will see how this works in practice in a moment.

The beginning of the projection is shown in Fig 48, which also shows the shape of the astronomical triangle on a zenith centred diagram. Note that we keep the same viewpoint of the side of the celestial sphere, that is, from the west, as we had for the projections in the last chapter, so that on the zenith centred projection in Fig 48 the pole still appears to the left of the zenith as it did in the pole centred diagrams.

When we project the observer's frame of reference, the altitudes and azimuths, we will obtain straight lines of azimuth radiating from Z, and

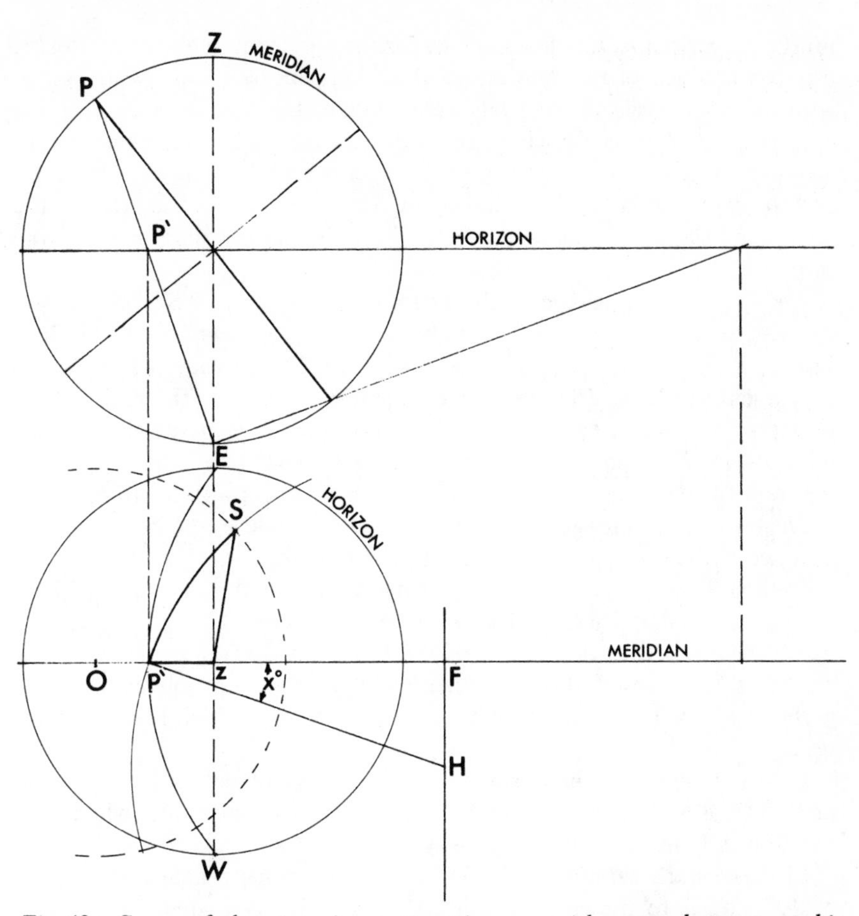

Fig 48 Some of the steps in constructing a zenith centred stereographic projection. Note that the construction for the star's Dec circle is not shown, but it is the same as for the almucantars of altitude on the pole centred astrolabe. Compare the shape of the astronomical triangle ZPS on this zenith centred projection with that on the astrolabe, (pole centred).

concentric circles of altitude around Z. On the other hand, in the zenith centred diagram the celestial co-ordinates will be distorted in the same way as were the almucantars of altitude and the eccentric circles of azimuth in the pole centred diagrams.

In Fig 48 the star, S, must be fixed on the projection by the intersection of two lines showing either the local or the celestial co-ordinates. In this case, three of the co-ordinates have been shown, hour angle, Dec, and azimuth. The construction of the declination circle has not been included on the diagram, but it is the same as for the almucantars in the astrolabe. The construction of the hour circle which happens to be east west (that is, with an hour angle of 90°) has been shown, together with part of the construction of the hour circle through the star. This hour circle makes an angle of $x°$ with the east west hour circle where it intersects at P′, and the centre at H can be found by the same method we used to find the centres of the circles of azimuth in the pole centred diagrams for the astrolabe.

Fig 48 shows the shape of the astronomical triangle P′SZ on a zenith centred diagram. It has two straight sides, and a curved side P′S. In a pole centred diagram, the triangle would also have a curved side, but in this case it would be ZS. The choice of centre, the pole or the zenith, must therefore be made according to which co-ordinates we are given and wish to draw to scale on the diagram. We do not want to draw the whole diagram for each problem we tackle, but by looking a little closer at the geometry of the stereographic projection we can find a number of simplifying constructions to cut the work to a minimum.

In the following examples we will solve some problems and develop the techniques to simplify the calculations and drawings as we go along. You would be advised to draw each of the following diagrams and try a few examples for yourself. You can also invent your own problems using the globe, which will give you a reasonable answer with which to check your result.

Let us assume that there is a star (or any other object) at Dec 24°, and of known RA. It might be the planet Mercury, for example, which you can find only in a bright twilight sky.

Find the star's azimuth and hour angle when it has reached an altitude of 40°, and is to the east of south, that is before it is due south. Your latitude is 51° north.

Which diagram shall we use? We know the required zenith distance at altitude 40°: it is $90 - 40 = 50°$. The zenith distance will be a straight line on the zenith centred diagram, and the azimuth will be an angle between two straight lines, so it will be most convenient to put Z in the centre of our projection.

Draw a line across the middle of the page to represent the meridian. This is the line on which the pole, P, and the zenith, Z, are to be marked (see Fig 49). We will use a scale factor of 10, so that if we drew the horizon it would be 10 units in radius, but as we are dealing with distances from the zenith it is unnecessary to draw the horizon. On the meridian, mark the zenith at Z. Now the pole will be to the north of the zenith, that is to the left on the diagram, and since the altitude of the pole is the same as our latitude, its zenith distance will be $90 - 51 = 39°$. Now we must use the stereographic scale. The actual distance from Z to P on our diagram will be $10 \times \text{Tan}\ \frac{39}{2}$, which is 10 Tan 19·5, which is 3·54 units. Now we could just as easily have chosen 30, 20, or any other value as our basic radius, when the length of this line on the diagram would have been, for example, 20 Tan 19·5, which comes to 7·08 units. But, using our base of 10 units, we mark P at a distance of 3·54cm from Z. (In these examples we will use centimetres throughout, but we could equally work in inches.)

The star has a zenith distance of 50°, which will be represented by a circle centred at Z of radius 10 Tan 25, since although its zenith distance will change all the time, we know that this is the value for the moment with which we are concerned.

Although we do not yet know exactly in which direction the star will be, we do have a rough idea, so that it is unnecessary to draw the whole circle. In Fig 49 the arc of this circle at about south east has been drawn with a solid line, and this is all that is strictly required, but the remainder of the circle has been shown as a broken line for this example to give the complete picture. This also applies to the other lines drawn in the figure. Those which are broken are given only to illustrate the complete construction which you can omit when you understand how it all works.

The zenith distance circle has a radius of 10 Tan 25 = 4·66cm and is drawn about Z. Now we must draw part of the declination circle. This is centred at some point O on the meridian as it was in Fig 48. What do we know about this circle? It will cross the meridian at two points. One will be south of the pole; the other will be on the meridian to the north of the pole. We can find both these points by calculating their zenith distances at the time; then the centre, O, will be halfway between them. Again, however, we do not need to mark the northern meridian transit point, since once we know its distance from the zenith, and have worked out and marked the southerly point, we can work out the radius and mark O at this distance from the southerly point.

At the southerly transit, the zenith distance of a star will be its co-declination (polar distance) minus the pole to zenith distance. In this example ZD = 66 − 39 = 27°. On our diagram this will be 10 Tan 13·5 = 2·4cm, so M can now be marked at this length from Z. At the northern transit, the zenith distance will be the co-declination plus the pole to zenith distance. This is 66 + 39 = 105°, which on our diagram would be 10 Tan 52·5 = 13·03cm. The diameter of the circle will therefore be the sum of the two zenith distances, 2·4 + 13·03 = 15·43cm. The radius will be half this, 7·71cm, and this is the distance to the centre O from the point M. We can now draw an arc from O of radius OM which crosses

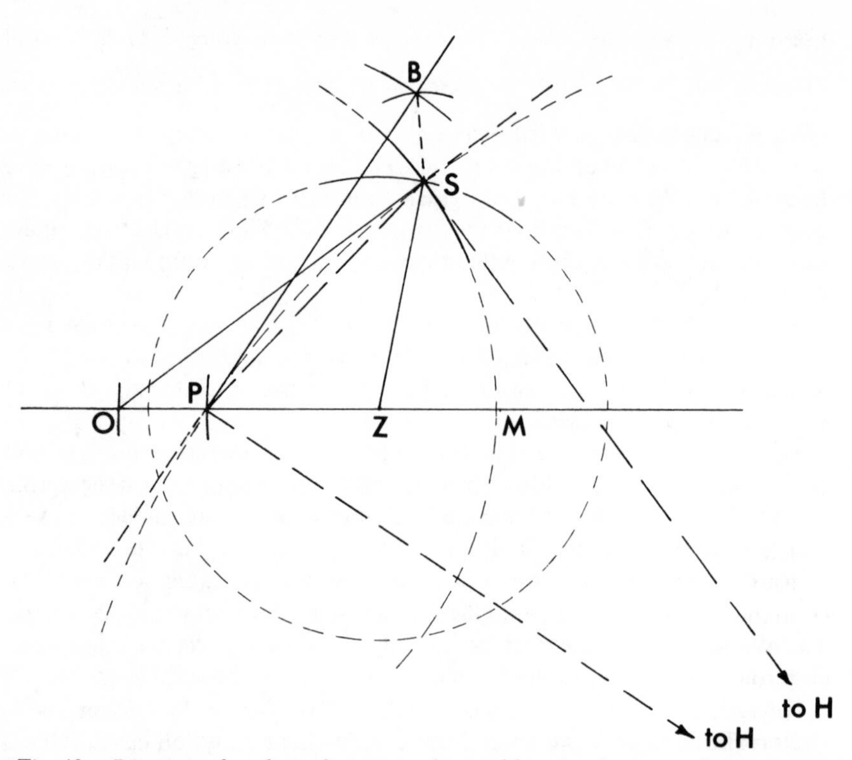

Fig 49 Diagram for the solution to the problem in the text, showing some of the construction lines (needed for proof only) as broken lines. Can you see how this diagram evolves from that in Fig 48?

the zenith distance circle at S, fixing the star's position. The two circles also intersect to the west of the meridian, but we are interested only in the easterly position.

We can now draw the line ZS, one of the straight sides (in this case) of the astronomical triangle, and measure the angle it makes with the meridian from the north point, the angle PZS, which is 102°. This is the star's azimuth.

The third side of the astronomical triangle, PS, is an arc. To find the hour angle we need to know the angle this arc makes with the meridian as it crosses it at P. Imagine that the required arc is centred at some unknown point H and has been drawn on the diagram, although it will not be necessary to actually draw it, so it is shown as a broken line in Fig 49. The arc from centre H must pass through P and S so that if we could draw the radii to these two points from H we could then draw tangents to the circle at these points also. The tangent at P would make the required angle with the meridian. But OS is also a tangent to the circle about H, since each hour circle will cross each circle of declination at right angles. We should now draw OS.

There will be a position of B on PB where OP = BS. Then we have two triangles with the common base PS, these are triangles BSP and OPS in which the lines OS and PB are both tangents to the same circle. It can be shown that the angles between two tangents and the chord of the circle common to both (the line PS) must be equal. So that if we make triangles OSP and BSP identical, PB will be a tangent to the circle to which OS is also a tangent, which is just what we want.

We have already have a common side PS, so we must make the other sides equal. SB must be the same length as OP, so we draw a little arc of radius OP from S. PB must be the same length as OS, so we draw an arc of length OS from the centre P. These two arcs intersect at B, and we can draw the tangent PB at the point P and measure the hour angle BPZ, which in this case is −56° (the minus sign indicating that the HA is measured eastwards from the meridian).

So the star S is at azimuth 102° and hour angle −56°, when it is at altitude 40°. We will now summarise the construction alone for this calculation, and you will see that it is not at all complicated compared with the full proof which we have just worked out.

Draw the meridian and mark Z on it. Mark P at the pole-zenith distance appropriate to the latitude, remembering to use the scale factor of $r \text{ Tan } \frac{\theta}{2}$, if possible using a value of 10cm or 10in for *r*. Draw an arc of the altitude circle at the approximate position of the star, with a radius equal to the zenith distance (that is 50°) from Z, to the same scale. Now work out the diameter of the Dec circle from the zenith distance at the north and south transits (converting both these values to the scaled lengths), divide by 2 to find the radius, and thus find the centre of the Dec circle from M, the south transit. Where this Dec circle intersects the altitude arc is the position of the star, S. The angle of the line from Z to S with the northern meridian is the azimuth. From S draw an arc of radius OP. From P draw an arc of radius OS. Connect the point where they intersect to P, and the angle this line makes with the meridian to the south is the hour angle.

The 56° hour angle we have just found can be converted to time units at the rate of 15° per hour, and 4 minutes per degree. Thus, −56° becomes $\frac{56}{15}$ = 3 hours, and the remainder, 11°, is 11 × 4 = 44 minutes, giving −3h 44m. The sidereal time is given by ST = RA + HA (see page 50), but since the star is to the east, HA is negative, so subtract 3h 44m from the RA to find the sidereal time at the moment in which we are interested, and then convert to mean time.

Now we will tackle another problem, but in less detail than the last. A few of the construction lines are still retained in the diagram, Fig 50, but even these could be left out when you are more practised. A star is observed at an altitude of 26°, and azimuth 140°, from latitude 51°N, longitude 0°, at 19h 42m on 1977 February 2. What is its RA and Dec?

In a zenith centred diagram we could draw a straight line from Z at the azimuth given, and the zenith distance circle. But as we do not know the Dec, we cannot find the eccentric centre of this circle, so then we are stuck. On a pole centred diagram, on the other hand, we could draw the almucantar for the star, we know the tangent to the azimuth circle and we know the Dec circle is concentric with the pole, so this sounds much more hopeful.

Once more draw the meridian, mark the pole, and then mark the zenith at the same polar distance as in the last example, 39° = 10 Tan 19·5 = 3·54cm. The 26° almucantar has a zenith distance of 64°. When it crosses the meridian to the south, the polar distance will be 64 + 39 = 103°,

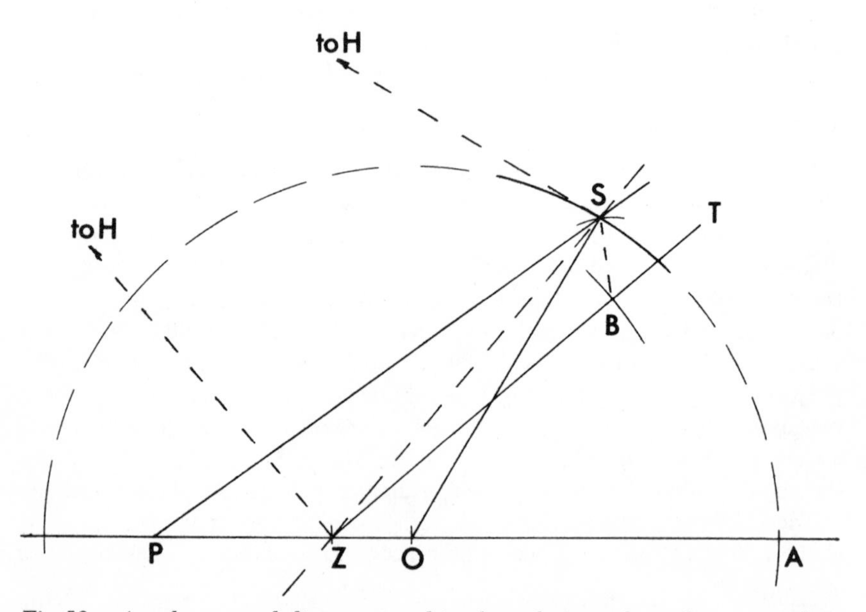

Fig 50 A pole centred diagram used in the solution of another example in the text. Fewer construction lines are included this time. Can you see which lines were drawn on the astrolabe in the last chapter?

which is 10 Tan 51·5 = 12·55cm from P. To the north of P, the polar distance will be 64 – 39, or 10 Tan 12·5, which is 2·21cm. The diameter of the circle will therefore be 12·55 + 2·21 = 14·76cm, so that the radius is 7·38cm. Point A on the circle is marked at 12·55cm from P (to the south), and the centre of the almucantar is 7·38cm north of A, which gives the point O. Once again, draw an arc at about the expected position of the star, S, with a radius of OA, and centre O.

This time, the arc of the circle of azimuth which would be centred somewhere at H has not been shown in the diagram, Fig 50. We know the angle which a tangent to this circle will make at Z with the meridian: it is the azimuth, 140°, measured as always from the north point. Draw ZT so that angle PZT = 140°. If we knew where S was, we could once more construct identical triangles on the common base ZS, with ZT and OS as tangents to that same circle centred at H. Side SB is the same length as side OZ, and OS equals ZB. But OS = OA, so we can find B on the line ZT by striking an arc of length OA from Z to intersect with ZT at B. From B we have only to strike an arc of length ZO and where it intersects with the almucantar is the position of S. Measure the length of PS. Since this is a point on the concentric Dec circle about P, it will have the same co-declination on the meridian, and we find that this is 11cm.

So $10 \text{ Tan } \frac{\text{co-dec}}{2} = 11$. The unknown angle divided by two has a tangent of 1·1, so it is 47·75°. The co-declination is twice this; 95°30′, and so the star is just below the equator, at declination –5°30′.

The angle between PS and the meridian is the hour angle to the east, and measuring angle SPA gives its value, –35°. Hence the hour angle is –2h 20m. Now we can calculate the sidereal time, but this must be done as accurately as possible. Adding the various times given in the table on page 194 for 1977 February 2, 19h 42m gives:

8h 47·1m + 1m + 3·3m + 19h 42m = 28h 33·4m. Reject 24h, gives ST = 4h 33·4m. Our longitude is the same as Greenwich, so there is no correction to apply for this. Since RA = ST – HA, we have RA = 4h 33·4m – (–2h 20m) = 6h 53·4m. The position of the star is therefore RA 6h 53·4m, Dec –5°30′.

Navigation by the stars

Almanacs give the Greenwich hour angle and Dec through the year for the Sun, Moon, and the most prominent stars and planets. This enables the navigator to work out the altitude of any of these bodies at a particular time from his assumed position, and compare them with his measured values to find the true position. Modern navigation methods include radio bearings, of course, but we can do with a home made quadrant and a pencil and paper what would otherwise need very expensive navigational equipment, and with an accuracy that is not much worse.

For navigational purposes, a compass bearing on a celestial object cannot be taken with sufficient accuracy to use as a precise position fixing method. The Sun's altitude can be measured with high accuracy with a sextant, or even our quadrant, but in this case a single altitude figure is insufficient to give our position. We need either two or more Sun sights separated by a few hours, or two or more star sights at about the same moment. The more sights we have, the greater the accuracy of our calculated position.

We will assume that, using the globe as in the example on page 90, we have estimated our position at latitude +38°, longitude 14° east. This is the equivalent of the mariner's dead reckoning position. The date is 1977 August 5, and from the almanac, or by calculations such as we will examine in the next chapter, we find that the Sun's position is RA 9h 01·5m, Dec +16°56′. With the quadrant we make an altitude measurement of the Sun at 14h 00m UT, when we find it is 48·3°, and again at 16h 00m, when we measure it as 24·6°. We can now draw two zenith centred diagrams, and since most of the calculations needed to draw these diagrams and much of the construction itself is common to

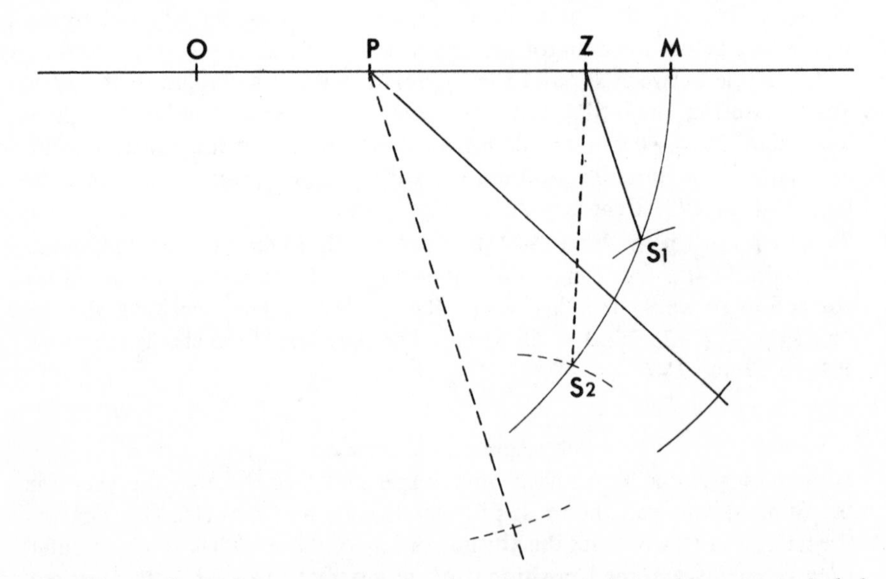

Fig 51 Only the lines necessary for the solution to the problem are included in this diagram, which is centred on the zenith.

both, we can draw them on the same chart.

Fig 51 shows the two diagrams, using solid lines for the 14h 00m calculation and broken lines for 16h 00m. This time all the lines not relevant to the problem have been left out and you can see how much simpler the diagram becomes, even containing *two* calculations.

If we were exactly at the assumed position, what would be the altitude and azimuth of the Sun at each of the two times we made the altitude measurements? We find these from a zenith centred diagram.

The distance from the pole to the zenith is, as always, our co-latitude, which in this case is 52°. On the diagram this will be represented by a spacing on the meridian of 10 Tan 26, giving 4·87cm. Now we know the Sun's Dec, so we can calculate its zenith distance when it is on the meridian both south and north of the pole. When the almucantar for co-declination 73°04′ crosses the meridian south of the pole, its zenith distance will be 73°04′ − 52° = 21°04′. On our diagram, this will be the point M at a distance from Z of 10 Tan 10°32′, which is 1·86cm. The zenith distance to the north will be 73°04′ + 52° = 125°04′, and on the diagram this would be 10 Tan 62°32′ from Z, or 19·23cm. The diameter of the almucantar is therefore 19·23 + 1·86 = 21·09 units, and the centre at O will be half this distance from M, 10·54cm. So now we can draw the meridian with P, Z, M and O marked on it and draw part of the almucantar to the south west where we expect the Sun to be; this is an arc through M at centre O.

To find the position of the Sun on this arc at the two times in which we are interested we need to know the hour angles, which can be found from the Sun's RA and the sidereal time. From the table in the appendix we find that at 14h 00m on this date, the sidereal clock is a total of 20h 55·9m fast on UT, so the sidereal time is 10h 55·9m at Greenwich. But we are 14° east of Greenwich, so our sidereal time will be 14 × 4 minutes fast on Greenwich, giving our local sidereal time as 11h 51·9m. The hour angle is the difference between the Sun's RA and the RA on the meridian, which is 11h 51·9m − 9h 01·5m = 2h 50·4m, or 42·6°. Two hours later at 16h 00m UT, the hour angle at our longitude will be 30° greater, 72·6°.

Now we can draw lines from the pole P at these two hour angles, and since it is past noon, and since the hour angle is positive, the Sun is to the west of the meridian in both cases. With P as the centre, mark an arc of radius OM across both lines. Now with these intersections as the centres, mark arcs of radius OP to cross the almucantar at S_1 and S_2, the positions of the Sun at 14h 00m and 16h 00m respectively. The azimuth of the Sun at 14h 00m is the angle MZS_1, 72·1° west of south, and at 16h 00m, the azimuth at S_2 is 86·7° west of north. From the diagram we measure the zenith distances of the Sun at both times, using the stereographic scale. S_1 is 3·88cm from Z, so 10 Tan $\frac{\theta}{2}$ = 3·88, giving θ = 42·5°. So the altitude of the Sun at 14h 00m is 47·5°. The length ZS_2 is 6·48cm, giving an altitude of 24°.

From our actual position, the measured altitudes of the Sun at both times were greater than they were at the estimated position. So in both cases we were nearer the Sun than our estimated position. To find the true position, we draw a very large scale chart of the area, marking the lines of longitude and latitude, and then plotting the Sun's position from the estimated position. The chart is shown in Fig 52. Draw the lines of 14° longitude and 38° latitude through the estimated position P. Now draw other lines of latitude parallel to the one through P, separated by a suitable scale distance *x*, which in this case represents 10′ of latitude.

Since lines of longitude converge the further north or south of the equator we are, we need the relationship between the two co-ordinates. This is exactly the same as for the lengths of celestial arcs, which was shown in Fig 45. A given distance representing a particular interval of longitude divided by the distance representing the same interval of latitude is the cosine of the latitude. From Fig 52 we could say that $\frac{y}{x} = \text{Cos (lat)}°$.

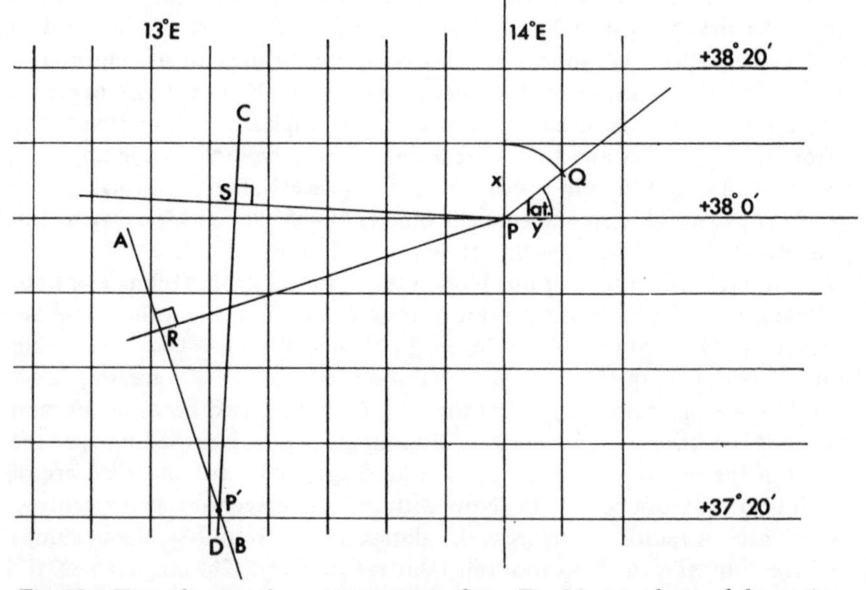

Fig 52 Transferring the measurements from Fig 51 to a chart of the region around our assumed position we can find the position line AB. We are somewhere on that line. A second calculation gives the position line CD. Where the position lines intersect at P′ must be our position.

This gives one way to work out the separation of the lines of longitude on our diagram. Another way involves a simple construction. Having drawn lines of latitude, put your compasses at P with radius x, the separation of the latitude lines for a given interval (in this case, 10′), and draw an arc. Now draw a straight line from P at an angle to the line of latitude through P equal to the latitude (in this case, 38°). Where this crosses the arc, at Q, drop a line of longitude, perpendicular to the latitude lines. Its distance y from the longitude line through P is the separation you need.

Having plotted a few lines of longitude and latitude around your estimated position, draw a line from P for the azimuth of the Sun at the first time, which on the diagram is the line PR at azimuth 72.1° west of south (the azimuth from the north point is, of course, 252·1°). Now you must move along the line until the Sun is at the measured altitude 48·3°. Since the altitude was 47·5° at P, you must be 0·8° closer to the Sun, that is 48′, from P. Using the *latitude* scale, mark this position, R, on the line.

At any instant of any day, there is a point on the Earth at which the Sun is exactly overhead. Around this point there are a number of places from which the Sun will appear to be at a particular altitude at the same instant, and all these places will be on a circle centred on the spot where the Sun is overhead. Of course, there are any number of such circles of different radii representing all the altitudes from 0° to 90°. These circles are mostly of such large radii, that on any detailed chart (except where the Sun is almost overhead) the circumference will appear as a virtually straight line. In Fig 52 we have an example of such a line; it is the line AB through R at right angles to PR. We know that at R the Sun will be at the altitude we measured, but so it will be on any part of the line AB. Our position is anywhere on this line, and it is therefore called a *position line*.

This is why we need our second Sun sight. Using the results of the sight at 16h 00m, we plot the Sun's azimuth, the line PS, and mark along it the difference between the calculated and measured altitudes, 36′. This gives the second position line, CD, and where this intersects AB is our corrected position, P′. This intersection of the position lines is called *a fix*. Our final position is, therefore, latitude 37° 21′ north, longitude 13° 11.2′ east.

We have worked through this problem in some detail, but once you have worked out a few examples for yourself you will be able to skip many of the detailed steps given above. As an alternative to the graphical methods described, you can solve the astronomical triangle mathematically, using the expressions given in the appendix. However, the graphical methods have the advantage of following the actual patterns in the sky, and so it is usually easier to spot any mistakes.

A Summary

Finally, here is a brief summary of the methods of 'astrographic' solutions to problems with a few extra hints.

It helps to avoid mistakes if you draw a small circle around the centre of the projection, for example, the pole in a pole centred diagram. The centre is involved little in the construction itself, but is usually the point from which the linear measurements of the radial lines and concentric circles are made at the end of a problem. Taking a zenith centred diagram as an example, the following are the usual stages:

1 Draw the meridian and mark the pole to zenith distance (co-lat) drawn to the scale $r \text{ Tan } \frac{\theta}{2}$, where r is any convenient value (eg 10cm) and θ is the angle.

2 Work out the south transit and north transit polar distances (or zenith distances in a pole centred diagram) and draw the almucantar at the radius so obtained, marking its centre, over an arc in the region of the diagram where you will be working.

3 Construct tangents to the circles of RA at P. The distance along this line from P to one point on the two congruent triangles drawn for this construction is the radius of the almucantar, and from this point, the star is at the intersection of an arc of radius OP, (ie the centre of the almucantar to the pole) and the almucantar itself.

4 Remember to construct this tangent at P when Z is the centre and vice versa.

5 Measure zenith distances in straight lines from the zenith in a zenith centred diagram, converting the lengths to an angle according to the $\text{Tan } \frac{\theta}{2}$ scale.

6 The azimuth is the angle at Z between the line to the star and the meridian.

7 Measure the angles at the pole (ie the 'non-centre') by the tangent at this point.

One last hint: when you are striking an arc from a point and it is of small radius so that it intersects with another line at two points close together, the intersection you want is the one which forms those two constructional triangles (for example, triangles OPS and BSP in Fig 49).

In many cases you will be able to tell in advance that the PZS triangle, the astronomical triangle, will be small on your diagram. For example, at moderate latitudes problems involving stars of high declination, particularly near the meridian, will give a small triangle, so in such cases you can decide from the outset to double the value of r in the stereographic scale.

EXERCISES ON CHAPTER 7

1 A star is observed through a gap in the clouds at 3h 00m on January 1, from latitude 51°N, longitude 0°, at altitude 37° azimuth 220°. What is the RA and Dec of the star according to our observations? What is the star?

2 The Sun is roughly south west at Dec −10° and the altitude is measured as 16·5° on October 19. If your latitude is 52° and longitude 5°W, what is the time (UT)?

3 The celestial co-ordinates of Mizar are RA 13h 21·9m, Dec +55°11′. On 1978 May 2 at latitude 40°N, longitude 75°W what is its altitude and azimuth at 21h 00m Eastern Standard Time?

4 From a dead reckoning position of latitude 42°5′N, longitude 11°42′W on 1978 November 1, two star sights are taken just before sunrise. Aldebaran (RA 4h 33m, Dec +16°25′) is measured at 7h 25m 30s UT at altitude 25°28′. Regulus (RA 10h 5·7m, Dec +12°13′) is measured at 7h 26m 15s at altitude 57°30′. What is your true position?

8
The Wandering Stars

The ancient Greek astronomers devised an ingenious picture of the Universe in which the planets were given special treatment. The fixed stars, it could be seen, turned about the Earth once a day, or very nearly. The Sun also went round the Earth, but since it changed its position

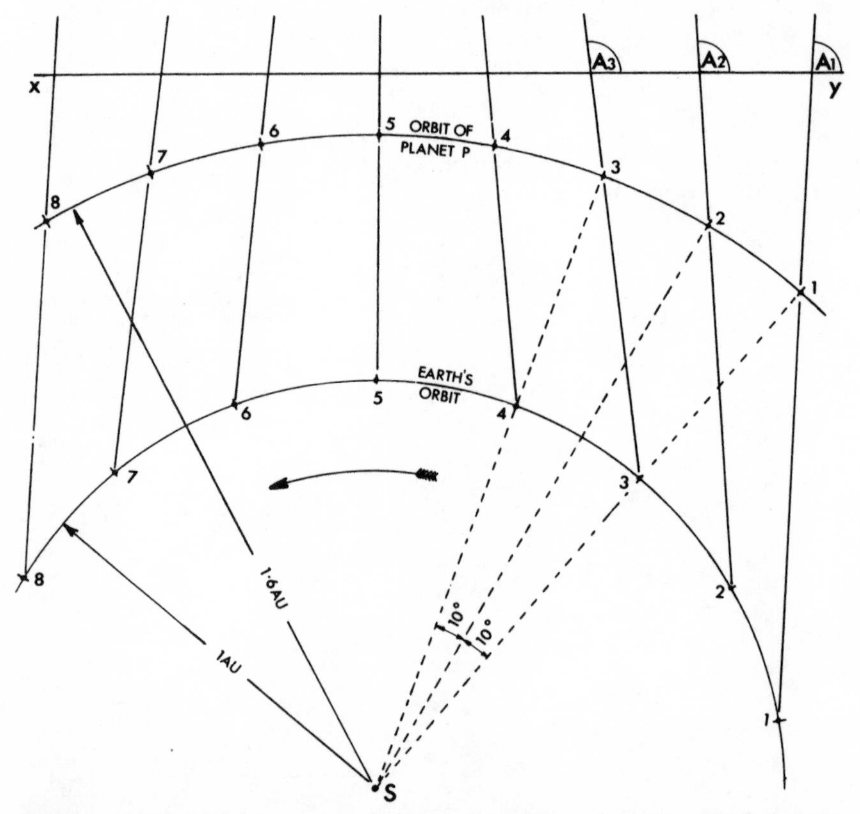

Fig 53 (Left) Motion of the Earth and a supposed planet which is at a distance of 1·6 AU from the Sun. The angles A_1, A_2, etc. are made with an east west line XY to show the relative position of the planet in the sky from

relative to the stars it must be moving about the Earth in a separate orbit to the stars. The Moon and the planets display this same independence, so they were also placed into separate orbits about the Earth.

The idea that the Earth was at the centre of the Universe and that the planets moved in circles about the Earth was fundamental, since to the Greeks, or at least to the followers of Pythagoras who was one of the most important of their teachers, the circle was a 'perfect' geometrical form. It is interesting to note that even an amateur ancient Greek astronomer could see with his own eyes that the planets did not go round the Earth in circles. The reason for this you too can see for yourself by watching a planet carefully for a few weeks. If the planets went round the Earth

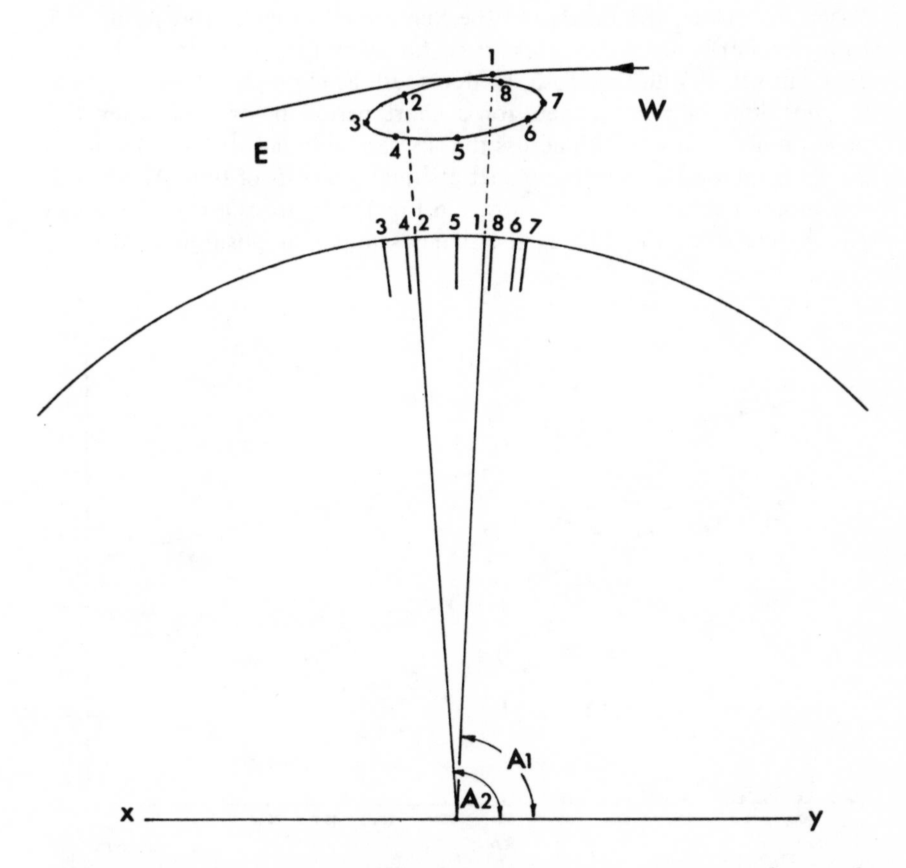

an observer on the earth. (Right), The angles A_1, A_2, etc projected on to the celestial sphere to show the planet's apparent path.

as the Moon does, even allowing for the fact that, as the Greeks knew, each planet goes at a different speed across the sky, they would describe a uniform path across the sky at a uniform rate.

This does not happen because the Earth is in motion as well. If we take the case of a fictional planet which is at a greater distance from the Sun than the Earth (a *superior* planet) and assume it moves around its orbit once in two Earth years, we can calculate its distance from the Sun as we will see in a moment. It would be about 1·6 times as far away from the Sun as the Earth. The planet would move around its orbit at half the angular speed of the Earth as measured at the Sun. Fig 53a shows the Earth and the hypothetical planet, at about the time that the planet is at its closest to the Earth, so that the Earth is almost between the planet and the Sun. When the planet, the Earth and the Sun are all in line, as at position 5, from the Earth the planet appears as far away in the sky from the Sun as it can get. It will, therefore, be due south at midnight. If we work out the positions of both planets for a short period before and after this position, its apparent path across the sky can be found. In Fig 53 a line is drawn from the Earth to the planet at equal intervals of time. Where this line meets the star sphere, as shown in Fig 53b by transferring the angles A_1, A_2 etc from Fig 53a, is the planet's apparent position in the sky

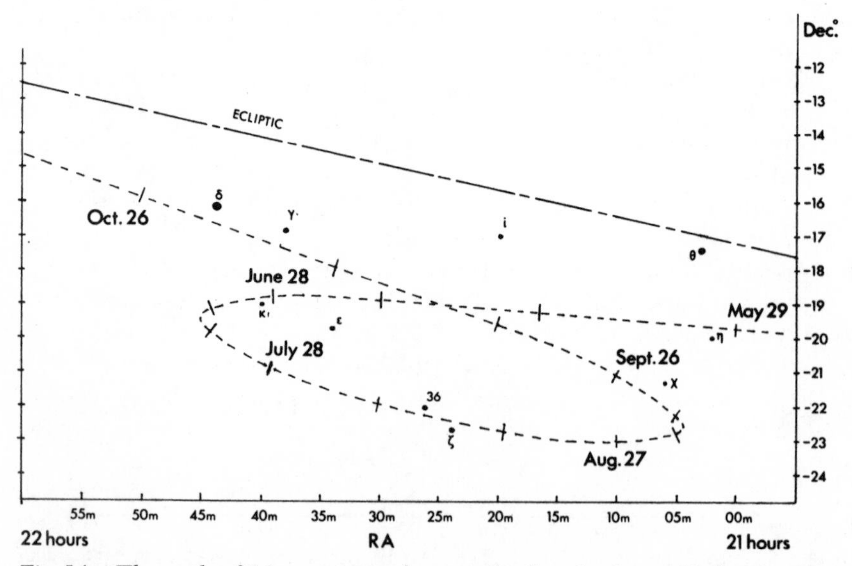

Fig 54 The path of Mars among the stars in Capricorn, at 10 day intervals during 1971, when the planet was at a very close opposition on August 10. Note the low, southerly Dec.

from Earth. We find that, due to the relative motion between the Earth and the planet, the planet appears to move backwards as the Earth overtakes it.

Once the Earth has moved a little more around its orbit, this *retrograde* motion of the planet ceases, and it once more resumes its easterly progress across the fixed star background, making a loop in its path as shown in Fig 53b. Fig 54 shows the path of Mars among the stars during part of 1971 during which it made a retrograde loop; its eastwards movement came to a halt, reversed, then the westward movement came to a halt and it resumed the easterly direction. The retrograde motion can also result in a zig-zag in the planet's path instead of a loop.

The fixation about circles did not disappear until after Copernicus had decided that the Sun, not the Earth, was the centre of the Universe. He found that circular orbits about the Sun could not explain all the slight variations in the planetary positions which could be observed. It was not until Tycho Brahe made many observations of planetary positions with an accuracy which is astonishing for one who had not got a telescope, that the 'breakthrough' was made. Tycho Brahe's observations of the planet Mars led Johannes Kepler, who worked with Tycho for some time, to discover that the planets orbit the Sun in elliptical paths. He formulated three laws, based on observations only, which Isaac Newton was later able to prove mathematically from his theories of gravitation and laws of motion. Kepler's laws are:

1 The planets move in elliptical orbits, with the Sun at one of the foci of each ellipse.
2 A line joining a planet to the Sun, called the radius vector, sweeps across equal areas in equal periods of time.
3 The square of the period of rotation of a planet about the Sun is proportional to the cube of its mean distance from the Sun.

It is from the third law that the distance of our imaginary planet with the 2-year period was calculated. The law holds good for all planets, so that we can say that the Earth rotates about the Sun in 1 year. The mean distance from the Earth to the Sun is called 1 *astronomical unit,* (abbreviated to AU) so that, using this unit, the constant of proportionality is the square of 1 divided by the cube of 1, that is 1. If then the period of a planet is T, and the mean distance to the Sun is *d,* we can say that $\frac{T^2}{d^3}=1$. Or, $T^2=d^3$, as long as we express the distance in AU and the period in years. The planet in Fig 53 has a period of 2 years, so that its distance in AU is $\sqrt[3]{T^2}=\sqrt[3]{4}$ $=1{\cdot}587$, close enough to 1·6 AU.

For the purposes of calculation it does not matter what the value of the AU is if all we want to know is where a planet is in its orbit at any particular time. All that is necessary is to know the relative positions of the Earth and the planet, and that is what we will tackle first.

The solar system

The Sun has a family of nine known major planets, all of which travel around their orbits in the same direction, and approximately in the same plane. This means that as seen from the Earth, none of the planets stray very far from the Sun's apparent path across the sky, the ecliptic. The distances involved are remarkable enough, but the differences between the distances of the different planets, particularly those outside the orbit of Mars are spectacular. For the moment we will look at the orbits of the Sun's four closest planets, which are, working from the Sun, Mercury, Venus, Earth, and Mars. Their mean distances from the Sun are 0·387, 0·723, 1, and 1·52AU respectively. Now from Kepler's third law you can work out the periods of rotation about the Sun.

The planets Mercury and Venus are both closer to the Sun than the Earth, so that as they move around their orbits it appears from the Earth that they never get very far away from the Sun. Mercury, in particular, can never be seen in the black night sky we might like to find it in. It is extremely elusive, but for northern observers it is easiest to see with the naked eye for a few days in the spring after sunset, and again in the autumn mornings for another short period. This is because the ecliptic is most steeply inclined to the horizon in the evenings around the spring equinox and mornings of the autumn equinox.

Venus can reach much greater angular separations from the Sun in the sky, and appears high in the sky at sunset and is visible well into the night on occasions, and similarly when it is a 'morning star'. This apparent misnomer, calling a planet a star, is used to describe the appearance of any of the planets to show whether they rise before the Sun, when they are morning stars, or afterwards, when they are evening stars. In the cases of Venus and Mercury there is never any doubt since these planets, called the *inferior* planets because they are closer to the Sun than the Earth, are always within about 50° and 30° from the Sun. The superior planets, on the other hand, can be exactly opposite the Sun when they are neither morning nor evening stars. When any of the superior planets are at this position on the sky, the difference between their RA and the RA of the Sun is 12h 00m, and they are at *opposition*. When a superior planet is at opposition it will be due south at about midnight.

When the RA of the planet and the Sun are the same, they are at *conjunction*, and the term is also applied when any astronomical body has the same RA as another body. The inferior planets line up with the Sun in two positions, one when the planet is closest to the Earth, and the other when the planet is at its greatest distance from the Earth. These are the *inferior and superior conjunctions,* respectively. Note that the superior planets cannot come between the Earth and Sun, so they can only be at conjunction (without the 'superior') on the opposite side of the Sun.

If we were able to look down on the solar system from a great distance above the northern hemisphere of the Earth, we would observe that all the planets move about the Sun anticlockwise. From our position on the Earth's surface, this means that the planets move across the celestial sphere from west to east. If we look at Fig 55, ignoring the construction lines, we can see the orbits of the Earth and Venus drawn to scale. The fact that they appear as circles, not ellipses, is hardly surprising. The difference in shape between an ellipse and a circle for these orbits, and those of nearly all the planets is very small indeed. This is an important point to remember for use a little later on.

In Fig 55 Venus, at V, and the Earth, at E, are both moving around their orbits anticlockwise. In accordance with Kepler's third law, Venus is moving faster than the Earth. A short time before the situation illustrated, when Venus was just a little closer to the Earth than in the diagram, the line EC through Venus would have been at a tangent to the circle of Venus' orbit. At this moment Venus would have appeared to be at its maximum angular distance from the Sun in our sky.

The angular distance of a planet from the Sun as seen from the Earth is called its *elongation*, to the east or west of the Sun. In Fig 55 the Earth's north pole is towards us, and the Earth appears to rotate on its axis anticlockwise (west to east). If you imagine that you are on the side facing the Sun, Venus has already passed overhead, so that it must be to the west of the Sun, near maximum western elongation. It will rise before the Sun as a magnificent morning star of magnitude −4.

Positions of the Planets

We can calculate very quickly the approximate position of a planet, just to see, for example, whether or not it will be visible on any particular night, or we can calculate this with some precision. In the process we learn a great deal about the orbits of the planets. We know how long the Earth takes to revolve about the Sun by measuring the Sun's progress around the ecliptic against the background of the celestial sphere. As a starting point we use the First Point of Aries which the Sun reaches at the spring equinox. This is the point through which we fix the 00h 00m hour circle, from which all the other values of RA are measured in an easterly direction around the celestial equator.

Since the planets, the Moon and the Sun are all to be found on, or close to the ecliptic, it is more convenient to refer their positions to the ecliptic rather than the equator. This can be done by measuring the angle between the planet and some starting point around the ecliptic, and by measuring its position above or below the ecliptic. This is called ecliptic longitude and latitude.

The *longitude* of a celestial object is the angle the object makes with the First Point of Aries, measured eastwards *along the ecliptic*. The *latitude* is the angle north (which has a positive sign) or south (negative) of the ecliptic, measured at right angles to the ecliptic.

Usually when we speak of celestial or ecliptic longitude and latitude we mean the angles relative to the ecliptic as seen from the Earth. But we can also use longitude and latitude relative to the centre of the Sun and clearly, since the Earth is in motion, its position must be taken into account when determining the position of the planets as we see them. Strictly speaking, the longitude and latitude should be called *geocentric longitude and latitude* when we talk about the apparent position from Earth. Otherwise, we must refer to *heliocentric longitude and latitude,* when the angles are measured round the Sun's centre.

If we know the heliocentric longitude and latitude of the Earth and another planet at a particular moment, and we know the periods of revolution about the Sun, called the sidereal period, P, we could work out the heliocentric positions for any other date. For the moment we ignore the angle the planet may be above or below the ecliptic, ie its latitude, because in most cases this is fairly small. We are also assuming, for the moment, that the planets move round their orbits in a uniform manner, which they do not.

As an example, let us take Venus. Given that the heliocentric longitude of the Earth at noon on 1972 January 1 was 100°15′, and that it has a mean daily motion around its orbit of 0·9856°, while Venus has a heliocentric longitude of 356°36′ at the same time, and its mean daily motion around its orbit is 1·602°, find the position of Venus in RA and Dec at noon on 1972 October 1. The mean distance of the Earth from the Sun is 1AU and the mean distance of Venus from the Sun is 0·723AU.

We can now draw a diagram of the two planets' orbits to scale, as in Fig 55. With the Sun at the centre, and assuming that the orbits are circular, we draw circles of radii 1 and 0·723 units; a typical measurement would be 10cm and 7·23cm. We mark the First Point of Aries with its symbol ♈, we can now mark the starting points for both planets. The longitudes must be measured eastwards from ♈. It is as well to adopt the same procedure for all these calculations, and to assume that the pole of the Earth in your own particular hemisphere is towards you in the diagram.

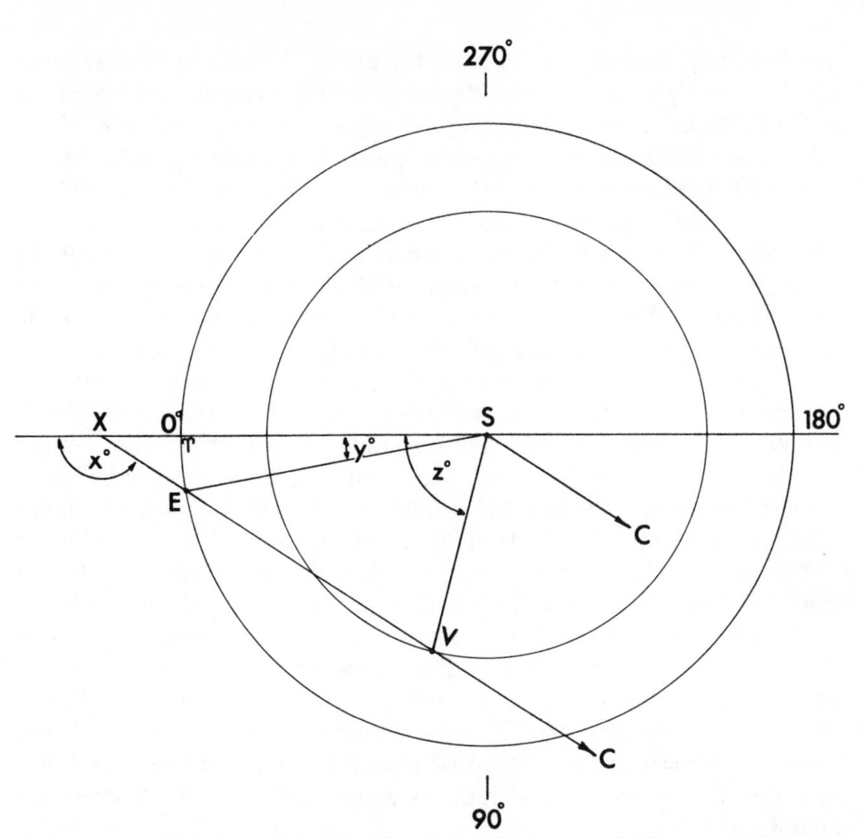

Fig 55 Simplified construction for the calculation of Venus' position on 1972 September 30, which assumes circular concentric orbits and uniform angular motion around the Sun.

Thus northern observers, for whom Fig 55 is drawn, measure the longitude anticlockwise. (Southern observers should measure clockwise.)

The next step is to count the number of days between the two dates in question; noon January 2 will be 1 day, noon January 3 two days and so on, giving 274 days to Obtober 1 (remember that 1972 was a leap year). Now multiply the mean daily motion of both the Earth and Venus by 274 to find the additional longitude that must be added to the January 1 figures. For the Earth, this comes to 270°, which must be added to 100·25°, giving 370·25°. This figure is larger than 360°, which means simply that during the interval in question, the Earth passed through the 0° position; so we subtract 360° to give the new longitude, which is 10·25°.

For Venus, the sum of mean daily motion for 274 days comes to 439°. Add this to the January 1 longitude and we have 795·6°. Venus has passed through the 0° position twice during the period in question, and we must reject all multiples of 360° (ie subtract 2×360°), giving its new longitude as 75·6°. We now plot both the October 1 positions on the diagram.

We have obtained the heliocentric positions of Earth and Venus for October 1, but to find RA and Dec, which are positions in the sky as seen from the Earth, we must find the geocentric longitude. The Earth's heliocentric longitude has been shown as angle y in Fig 55, and Venus's at angle z. From the Earth, E, Venus at V appears to be in the direction of the line EC. We must remember that the celestial sphere is of infinite radius, so that the angle we need must be measured in the same direction as all the others we have measured; the Earth-Sun distance is zero in an infinite circle so that the centre is still S. In other words, the line SC drawn parallel to EC gives the geocentric longitude, the angle XSC. We need not draw the line SC, however, since, because it is parallel to EC, the angle x made by projecting the line EC to cross the 0°–180° line is the same as XSC. The geocentric longitude in this case is 147·6°.

To turn this figure into RA and Dec we could look at a map in a good star atlas which plots the ecliptic across the stars and marks the divisions of longitude. In case you have not yet acquired this important aid to practical work, the RA and Dec of the ecliptic can be worked out from Fig. 56. This shows only a quarter of the whole ecliptic, from longitude 0° to 90° against RA and Dec to give a reasonable scale for accuracy. The curve is symmetrical about the hour circles 00h, 6h, and 18h, so the RA and Dec of any value of ecliptic longitude can easily be worked out as follows:

For ecliptic longitudes from 0° to 90° read directly from the curve given.

For longitudes from 90° to 180° subtract the longitude from 180° and then read directly from curve, and subtract RA from 12h.

From 180° to 270°, subtract 180° from the longitude, read from curve, but add 12h to RA, and Dec is negative.

From 270° to 360°, subtract the longitude from 360°, read from curve, and subtract RA from 24h, and Dec is negative.

In case this seems unduly complicated, the reasons for these steps are summarised in the small insert diagram in Fig. 56.

In this example, we have geocentric longitude of 147·6°. So we find RA and Dec for 180°−147·6°=32·4°, giving RA 2h 4m, which subtracted from 12h gives 9h 56m, and the Dec is +12°40′. The ephemeris gives the position of Venus on this date as RA 9h 49m, Dec+13°05′ so our approximate method is within 2° of RA and 1° of Dec, which is quite adequate if we want to know where Venus will be on this morning; it is very close to the star Regulus in Leo. (You may remember that Regulus is very close to the ecliptic).

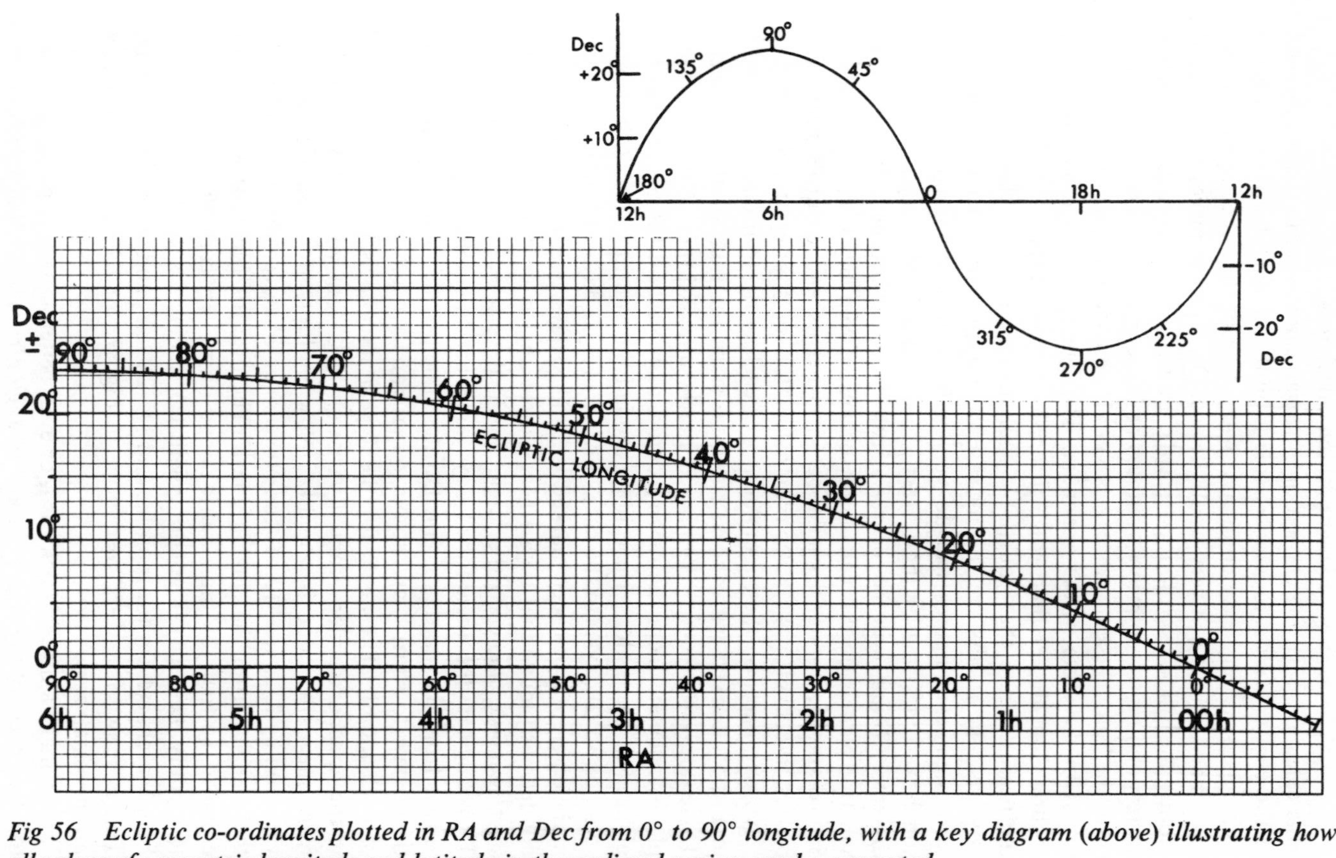

Fig 56 *Ecliptic co-ordinates plotted in RA and Dec from 0° to 90° longitude, with a key diagram (above) illustrating how all values of geocentric longitude and latitude in the zodiacal region can be converted.*

Planet position calculators

This calculation can be carried out mechanically using two discs pivoted at the centre, with a longitude scale around the outside. To make such a calculator is very easy compared with some of the other apparatus we have made, and it is a very useful device. On separate pieces of thin card (about the thickness of a postcard) draw four circles with diameters corresponding to the mean distances of Mercury, Venus, Earth and Mars from the Sun. These distances were given on page 140, and are also to be found in the table in the appendix. The figures in this table are to an accuracy far greater than you will need for most purposes; take them to the nearest figure that you think you can measure. For this calculator, however, the results are in any case only approximate so that the first three significant figures will do very well. Note that Mercury's orbit is some five times smaller than Mars', so if you give Mercury's disc a 5cm diameter, Mars' will be nearly 25cm. You cannot include the other superior planets on this calculator, because if Mercury's disc was made a 5cm diameter, Jupiter's would be 68cm across and Saturn's nearly $1\frac{1}{4}$ metres. We shall make a separate calculator for Jupiter and Saturn.

Make a mark on the circumference of each of the four circles as shown in Fig 57 marked with the astrological signs for each of these planets. These are ☿ Mercury, ♀ Venus, ⊕ Earth, and ♂ Mars. Now, for each planet calculate from the mean sidereal daily motion the angles that the planet will move along the ecliptic around the Sun in 1 day, 2 days, and so on up to 10 days. (In the case of Mars every 2 days will suffice; in fact, in the diagram it has been marked at 5 and 10 days, then the 5 day intervals divided into individual days.) Mark these angles anticlockwise from the starting mark, and also the angle for 20 days (twice that for 10 days). Then calculate the angle for 31 days and mark this 'February', measuring always from the starting point. Do the same for the 51 days to March 1, the 90 days to April 1, and so on so that you have a mark for the first day of each month. The angle for the 1 year interval is given in the table in the appendix under 'Annual variation in L', but you can work it out from the daily value if you like. Now multiply this figure of annual variation by 2, 3, 4, 5, 10, 20 and 50 years or whatever you like, and mark these angles from the starting mark. The disc for Mercury will be very crowded, so that you must mark the positions very carefully, and be sure to identify them as you go along with 1D, 2D etc for the daily variations, Feb, March, etc. for the months, and 1Y, 2Y, 3Y, and so on for the years. This is shown for Mars' disc alone in Fig 57, for the sake of simplicity on this small scale.

Now make a disc of thick card, or even better a piece of paper glued on to a thin sheet of plywood, marked in angles from 0° to 360° anticlockwise

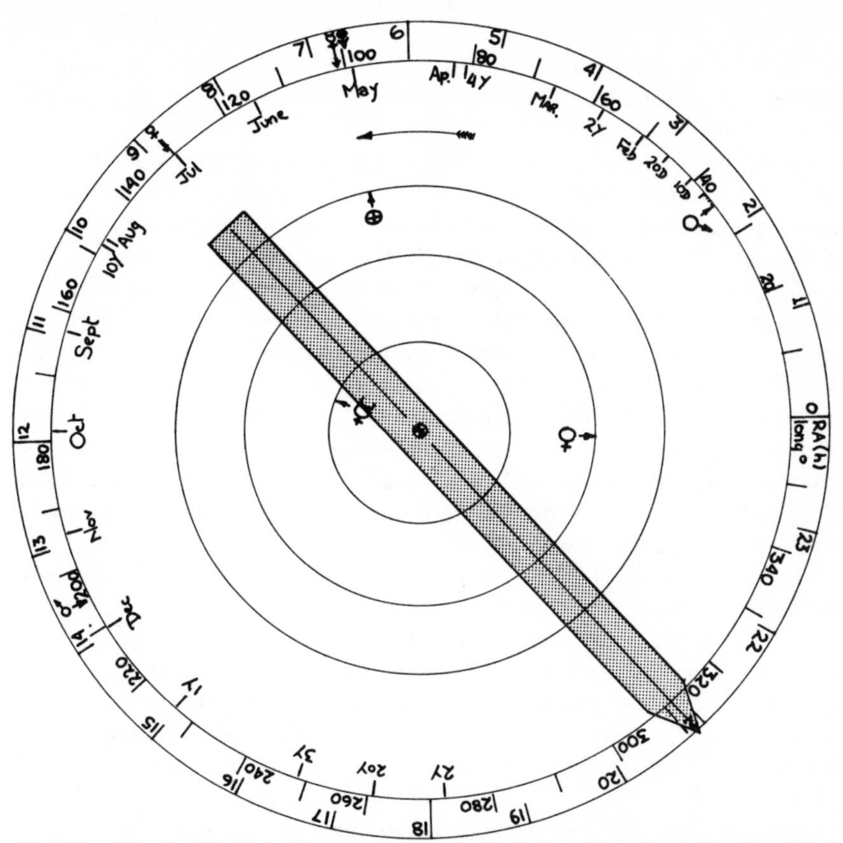

Fig 57 The finished planetary position calculator, based on the simplified technique, showing the marking of daily, monthly and yearly angles around the orbit of Mars only. The calculator is shown set for 1972 January 1.

for the heliocentric longitude L. From the graph in Fig 56 you should also mark the hours of RA subdivided into at least 20 minute intervals, marking each hour on the scale opposite its corresponding value of geocentric longitude. On this scale make an identifying mark, such as the astrological symbol, at the longitude of each of the four planets for 1971 Jan 0·5 from the table in the appendix. (Jan 0·5 means simply half a day after the beginning of the year, ie noon on January 1.)

Finally, cut out a strip of thin transparent plastic on which a line has been scratched down the middle and filled with ink as we did for the pointer on the astrolabe. At a distance from one end of this line equal to the outer radius of the RA/longitude scale, push through a drawing pin. Cut out the four circles, and push the pin first through the pointer, then Mercury's disc centre, then Venus, then Earth and last of all Mars; then pin them all to the centre of the thick disc with the longitude scale on it. If necessary, place a small piece of wood behind the centre to give the pin a secure hold.

To use the calculator, set the planets against their starting marks, the heliocentric longitudes for 1971 Jan 0·5, beginning with the outermost planet you are interested in (you will always have to position the Earth, of course). Place the line of the strip, which is called the 'cursor', on the number of days which have passed in the month you are interested in. These days are marked around the planet's disc. Hold the cursor in this position, and move the disc until the planet is under the line. Now repeat this procedure for the month and the year, and don't forget to add 1 day's extra motion for every leap year since 1971. If the particular number of days or years is not marked on the scale, it does not matter; simply make two or three movements which add up to the right number from those that are marked. Now set the other planet, the Earth, if Mars was the first you set, in the same way.

The calculator in Fig 57 is set up for all the planets on 1972 January 1 and it is shown ready to find the ecliptic longitude of Venus. The pointer is centred on the Sun, because we must read the scale at 'infinity', so it must be set parallel to the Earth-Venus line. You can either do this by eye, or use a ruler if you want to be precise. The pointer is set up in the diagram, and it indicates Venus's geocentric longitude as about 312°, and RA 20h 55m. At this stage we are still assuming that the planets are all on the ecliptic.

Using exactly the same constructional method you can now make a second calculator for Jupiter and Saturn, whose symbols are ♃ and ♄. The diameter of the Earth's orbit will need to be about 5 to 7cm rather like Mercury's in the first model. Even a 2cm radius will give a 19cm radius for Saturn, so there is not much point in trying to include Uranus' orbit on the model—that would make it over 1m across. These models certainly help to give an idea of the scale of the solar system. If Pluto were to be included on this scale, the model would be over 2m diameter.

The outermost planets move quite slowly each year. As you can see from the table in the appendix, even the fastest moving of the outer three, Uranus, moves only just over 4° around the Sun in a year.

It is instructive to draw a the orbits of the nine planets on a large sheet of paper, without bothering about the scale. Merely mark their heliocentric longitudes around their orbits at intervals up to 50 years from now. Although the resulting diagram cannot be used to show the geocentric longitudes, the planets' apparent positions in the sky, it gives a clear indication of what part of the sky each planet will be in, and a rough indication of the dates of conjunctions and oppositions.

Bodes' Law and other curiosities

A number of curious devices have been found to explain the distances of the planets from the Sun. Kepler himself, whose third law was confirmed by Newton, tried to explain the sizes of the planetary orbits by fitting solid geometrical figures inside the 'spheres' around which the orbits were fixed. Then a curious series of numbers was noted by Titius in 1772, and formulated by Bode in 1778. This 'law', in fact unrelated to any known law of celestial mechanics, now bears Bode's name. It is quite useful for remembering the distances of the planets.

Start with the series of numbers 0, 3, 6, 12, 24, 48, etc and add 4 to each giving 4, 7, 10, 16, 28, 52, etc, Now divided by 10, and this gives, the distance of the planets from the Sun in astronomical units thus

Bode:	0·4	0·7	1	1·6	2·8
Planet:	Mercury	Venus	Earth	Mars	Asteroids
Correct Distance:	0·39	0·72	1	1·5	—

Bode:	5·2	10	19·6	38·8	77·2
Planet:	Jupiter	Saturn	Uranus	Neptune	Pluto
Correct Distance:	5·2	9·5	19·2	30	39·4

It can be seen that the law begins to break down for Neptune and Pluto, and it appears that Neptune should be where Pluto is.

An even more curious coincidence (?) has been found. If the distances of the planets from the Sun in AU are Me, V, Ma, J, S, U, and N, respectively, then:

$$\frac{\mathrm{Me} + \mathrm{E}}{\mathrm{V} + \mathrm{Ma}} = \frac{\mathrm{J} + \mathrm{U}}{\mathrm{S} + \mathrm{N}}$$

and the result is almost as good if you substitute multiplication signs for the addition signs in the 'equation'.

Elliptical orbits

We have dealt at some length with the approximate methods of calculating the positions of the planets. Now we can refine the method a little. It will still be approximate, but the errors can be made gratifyingly small with only a little extra effort. You will need a set of trigonometrical tables for this work and, if you can use one, a slide rule reduces the chores of long multiplication; but log tables can also be used, and are not much more trouble.

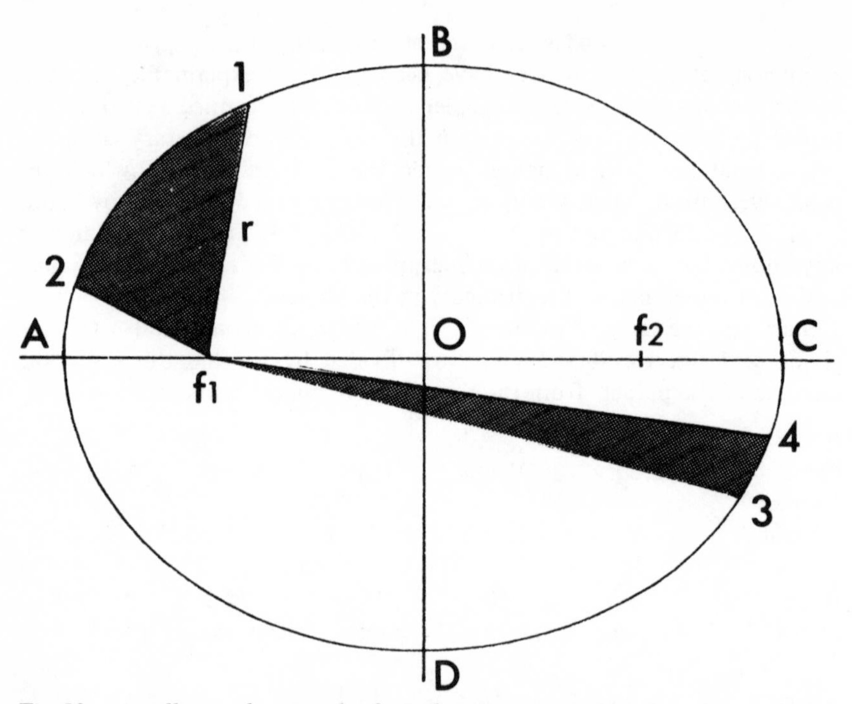

Fig 58 An ellipse, showing the foci, the major axis AC, the minor axis BD and two equal areas between one focus, f_1, and the points 1 and 2, and 3 and 4 to illustrate Kepler's second law, where r is the radius vector.

An ellipse is shown in Fig 58 to illustrate a planet's orbit. Kepler's first law says that the Sun is at one focus of the ellipse, in this case f_1. As the planet moves round its orbit it is at a distance *r* from the Sun. The length of r, which is the radius vector, is constantly changing, but it sweeps an area in the ellipse moving from point 1 to point 2 in time *t* days. At another part of the orbit, the radius vector sweeps an equal area while the planet moves from point 3 to point 4, and therefore, according to Kepler's second law, the planet takes the same time, t, to move from points 3 to 4 as from points 1 to 2. In general terms, this means that the planet moves faster along its orbit when closest to the Sun. It is moving at its fastest at the point A. This point of closest approach to the Sun is called *perihelion,* and the opposite point, where the planet is at its greatest distance is called *aphelion.* Each of these extremities of the major axis of an orbit is called an *apsis* (plural apsides) and the line joining them, or extended beyond them, is the line of the apsides.

The next factor to complicate matters is that the orbits of the other planets do not lie in exactly the same plane as the Earth's orbit; they are inclined to the ecliptic. This inclination, symbol *i,* is shown in Fig 59 which shows an elliptical orbit passing through the plane of the ecliptic. There are, of course, two points where the planet's orbit passes through the plane of the ecliptic, and these are called the *nodes*. The line joining these, through the Sun, is the line of the nodes. Now as the planet moves through one node it will move from below the ecliptic to above it. This is called the *ascending node,* for obvious reasons, and the other is the *descending node*. These are designated by the symbols ☊ and ☋ respectively and both must be exactly on the ecliptic.

In Fig 59 the angle from the First Point of Aries to ☊ is the longitude of the ascending node. Perihelion is marked at longitude ϖ at position P, and aphelion at A. We have now covered nearly all the elements of the planetary orbits which are listed in the table in the appendix, but one remains; the eccentricity. This is most easily explained by referring back to Fig 58. The centre of the ellipse is at O, but the sun is at f_1. The degree of eccentricity is given as the distance from the Sun to the centre as a

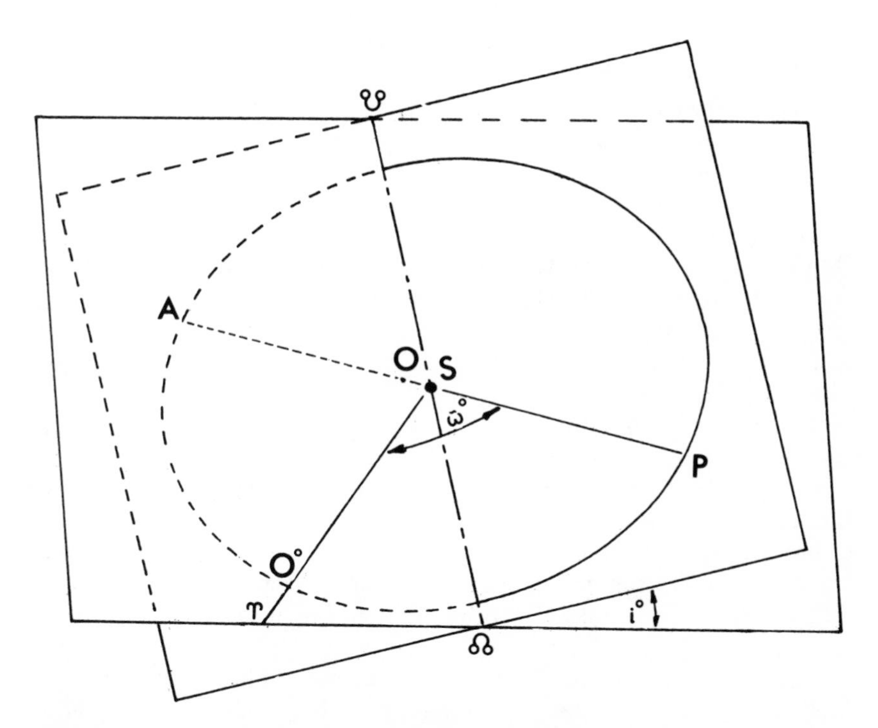

Fig 59 The main elements of an elliptical orbit, inclined to the ecliptic.

fraction of the semi-major axis of the ellipse. In this case it is $\frac{f_1O}{AO}$. This defines the shape of the orbit, and the actual size of the ellipse is immaterial.

In practice, the displacement of the Sun from the centre of the ellipse and the variations in the speed of a planet around its orbit have a much greater effect on positional calculations than the actual shape of the orbit. In fact, even the most elliptical orbits, those of Pluto and Mercury, are pretty good circles. Try drawing Pluto's orbit to scale (that is with the right value of *e*, the eccentricity).

Tie a piece of cotton into a loop 100mm long. Place two drawing pins 40mm apart into your paper along a line (AC in Fig 58). Now draw the ellipse by putting the loop over the pins and, holding the loop taut with your pencil inside it, pass the pencil around the pins. To the unaided eye this can just be seen to be a flattened circle. Its eccentricity is 0·25 very close to Pluto's orbit. If you repeat the test with the dimensions adjusted to give the respective eccentricities of the other planets, you will hardly be able to tell the difference from circles but you will be able to see that the Sun, which is one of the pins, is not at the centre.

So the first approximation of our more exact method is to assume that the planets have circular orbits, but are centred at some position on the line joining the perihelion to the aphelion at a distance from the Sun calculated from *e*. We multiply the planet's mean distance from the Sun by its eccentricity to find this separation of the centre of the orbit from the Sun.

We will continue to use AU and apply an appropriate scale to suit each problem. Fig 60 shows a scale drawing of the orbits of the Earth and Mars, with all the important points on the orbits shown, although the centres of the two circles O_e and O_m for the Earth and Mars are for our construction only. This is also the drawing we shall use to carry out calculations about Mars' position, so its construction must be carried out carefully as follows.

Draw a circle for Mars' orbit at the centre O_m as large as you can manage on the page. It will help to choose a radius which you can easily divide by 1·52, because this corresponds to Mars' distance from the Sun in AU and simplifies drawing the other orbits to the same scale. Draw a line $A_m P_m$ through O_m to represent the line of the apsides (perihelion to aphelion). We can now mark the Sun on the chart. The distance from O_m will be the radius of Mars' orbit multiplied by the eccentricity, 0·0934. If, for example, you have drawn Mars' orbit with a radius of 152mm, the distance from O_m to S will be 14·2mm. Since Mars is closest to the Sun at perihelion, P_m, the Sun will have to be between O_m and P_m. Now, we know the heliocentric longitude of the line from S to P_m; it is ϖ, the

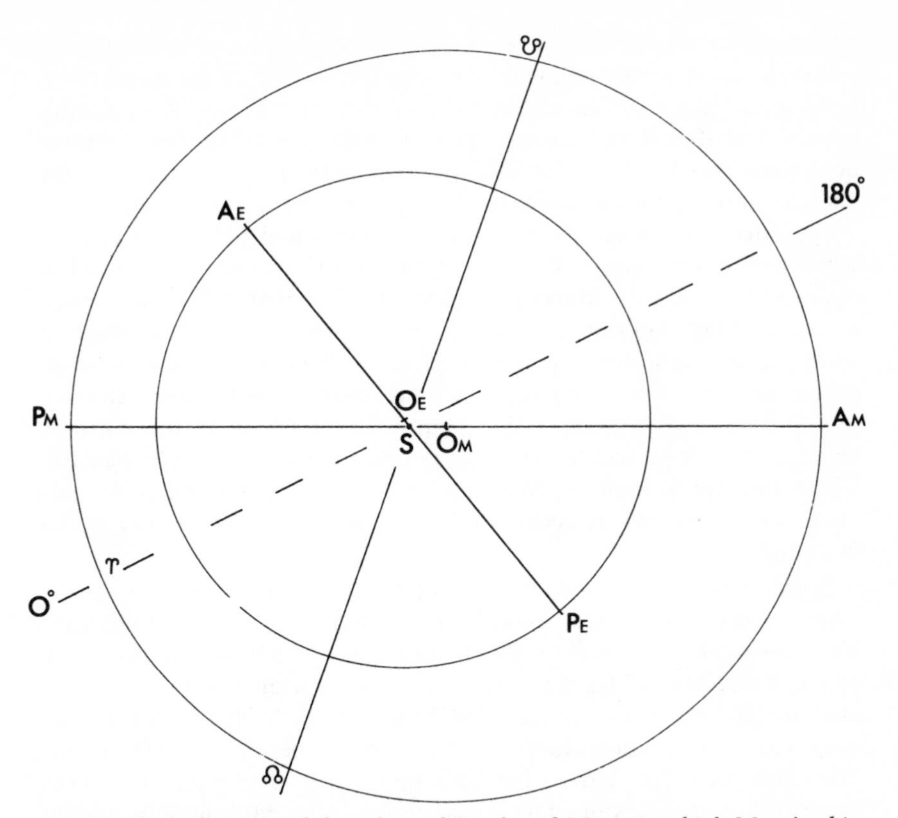

Fig 60 Scale drawing of the orbits of Earth and Mars, in which Mars' orbit is shown as a circle centered at 0_m*, and showing the positions of the perihelia and aphelia of the two planets, the line of the nodes for Mars, and* ♈ *(0° longitude*).

longitude of perihelion which, from our table, we find is 333·53°. The position of ♈ at O° will therefore be a little further around the Sun in an anticlockwise direction at an angle of 360°–333·53°. Mark this point and draw the 0°–180° longitude line through the Sun at S. This line will form the base for all our longitude measurements.

To complete Mars' orbit all that remains is to mark the nodes. We mark the ascending node at longitude 49·3° from the base line (anticlockwise as always) and mark ☊, then draw a line through S to give ☋.

Now we must find the centre of the Earth's orbit. We know that the centre O_e lies on the line through the Sun joining the perihelion and aphelion, so we plot the position of perihelion from the heliocentric longitude, given in the tables 102·44°, then draw the line P_e A_e. The centre will be on the same side of the Sun as aphelion, at a distance on our scale equal to the radius of the orbit multiplied by the eccentricity. If we used 152mm for the radius of Mars' orbit, the radius of the Earth's orbit will be 100mm and the distance from S to O_e will be 1·7mm. We can now draw the Earth's orbit at a radius of 100mm from O_e, and the diagram is ready for our calculations.

We can see from the diagram that the Earth and Mars vary in their separation enormously. Mars is very much closer to the Earth when opposition occurs at Mars' perihelion, at P_m, than when opposition occurs at Mars' aphelion. You can see as well, that the longitude at which these favourable oppositions occur is 333·5° and, since the heliocentric and geocentric longitudes are the same at opposition, reference to Fig 56 reveals the depressing reality for observers in the northern hemisphere that at such times Mars is well to the south of the equator. The maximum altitude of Mars as seen from London would be only about 24°; moreover, it occurs in August, when the evening sky is still fairly light.

Such an opposition of Mars took place in 1971 when Mars was very close to perihelion. Where were Earth and Mars at the next opposition? You can work this out from the difference in the annual variation in L of Earth and Mars. This is about 191·5°, which means that exactly 1 year after the last opposition Mars is 191·5° to the Earth, that is not far past conjunction. Two years later, the difference will be 383°, or 23°. so that Mars will have just passed the next opposition. The interval between successive oppositions of Mars is just under 2 years. If you now take the difference in the daily variation of longitude you can roughly estimate the interval between oppositions.

This period between successive oppositions of a planet is called the mean *synodic period*. It is a measure of the time the planet takes to return to any of the particular configurations with the Earth (opposition or, conjunction). The time a planet takes to make one 360° revolution about the Sun is called its *sidereal period*. For the Earth, the sidereal period is 365·256 days. By measuring the difference in the dates and times between successive oppositions (or say eastern elongations of the inferior planets)' we can calculate its sidereal period approximately.

Note the moment that one of the superior planets are on the meridian at midnight one year, and then observe when it next occurs. Convert the interval to years (of 365¼ days) then the sidereal period in years can be deduced from the synodic period in years thus:

$$\text{sidereal period} = \frac{\text{synodic period}}{\text{synodic period} \pm 1}$$

, (+for inferior and – superior planets). Since the value can differ quite a lot due to the changing velocity of the Earth and other planets at different places in their orbits, the value can vary according to the time of year. So it is best to try this experiment with Jupiter or Saturn first since these planets reach an easily observable opposition each year.

To find the positions of Earth and Mars on the orbits drawn in Fig 60 with reasonable precision, we must take into account the most disturbing influence in a planet's longitude, the *equation of the centre*. This measures the change in mean longitude due to the elliptical nature of the orbit. It is the main factor in the equation of time which is the deviation of the Sun from its mean geocentric longitude due to the changing speed of the Earth in its orbit.

The equation of the centre consists of a series of terms which should be applied to the mean position. Fortunately, the first term in the series is by far the biggest and for our purposes it will suffice: we will ignore the others. It will still allow us to find the planet's position to well within half a degree.

The equation of the centre depends upon the eccentricity of the orbit and the position of the planet relative to its perihelion. We can regard the perihelion of each planet as fixed on its orbit, although it does increase very slowly indeed, and the eccentricity is also fixed, so all we need is the mean longitude. The correction we apply to the mean longitude is:

$$2e \operatorname{Sin} (L - \varpi) \text{ radians.}$$

If we express this in degrees, using the usual $\frac{180}{\pi}$ conversion factor we can simplify to:

$$L' = L + 115e \operatorname{Sin} (L - \varpi) \quad \text{degrees.}$$

L' is the corrected value of longitude, and ϖ is the longitude of perihelion. The value $L - \varpi$ is called the mean anomaly of the planet. A table of the mean elements of the various factors you will need to know is given in the appendix. As well as the value of eccentricity, e, the table also gives the value of 115e already worked out.

Now we can use the equation of the centre to improve our planetary position calculation. As an example, what is the position of Mars in RA and Dec at 1975 July 15d 00h?

Just a word about counting the days. The positions of the planets and the orbital elements are given in the table for Jan 0·5 days. It is given in this form in astronomical ephemerides, so it is as well that you become accustomed to it. What the 0·5 means in fact is 12 hours after midnight January 0! In another 12 hours it will be 00h January 1. If we started from the Jan 0·0d position we could simply add one day for Jan 1 at 00h, two days for Jan 2 at 00h and so on. We shall do this for the Moon, since its mean elements are not normally quoted in ephemerides at the same date each year, so we can start when we like. But for the planets, to find the time at midnight we must finally subtract 0·5 days. Another important factor is that we must not neglect the motion of the planets during February 29. For each leap year we must add one day to our total.

In this example we have 182 days to July 1·5, to which we add 14 to take us to July 15·5 (total 196 so far), add one day for leap year 1972 and subtract the 0·5, giving 196·5 days.

We can start with the Earth. At 1971 Jan 0·5 its mean longitude we find from the table was 99·504°. The annual and daily change in longitude is given, so we need to know the total time to have elapsed since this starting date. To help with this the days of the year (not a leap year) have been counted to the first of each month in another table in the appendix. So we have four years' motion, $4 \times -0{\cdot}239° = -0{\cdot}956°$, and a total of 196·5 days giving $196{\cdot}5 \times 0{\cdot}956 = 193{\cdot}749°$. The mean longitude is therefore $99{\cdot}504 - 0{\cdot}956 + 193{\cdot}749 = 292{\cdot}297°$.

From Earth's mean position of 292·297° we can work out the magnitude and the sign (addition or subtraction) of the correction. It is 1·92 Sin (292·297 − 102·441). The Earth's anomaly is 189·856° so the sine is negative, and the correction must be subtracted. Its value is −0·328, giving the corrected longitude $L' = 291{\cdot}97°$.

We now repeat the procedure for Mars. The mean longitude works out to 352·09° and the anomaly is 16° 34′. The value of the equation of the centre, 10·7 Sin 16° 34′ is +3·05°, so Mars' corrected longitude is 355·14°.

We can now carefully mark Earth and Mars on the diagram, Fig 61, measuring the longitudes anticlockwise with the Sun as centre and with the First Point of Aries as zero. Draw a line from Earth through Mars and either draw a line parallel to this from the Sun and measure the longitude so obtained as explained earlier, or measure the geocentric longitude where the line from the Earth crosses the ♈—Sun line as shown. The geocentric longitude is $\lambda = 38{\cdot}5°$.

We see from the diagram that Mars has not yet reached its ascending node, and because its orbit is inclined to the ecliptic by nearly 2° it is south of the ecliptic. We can measure this ecliptic latitude by another small diagram to the same scale as the first, representing a side view of the ecliptic. Because the angles involved are so small we will multiply them by 10, find the latitude and divide by 10 to get the result. This introduces far less error than trying to work to scale with the correct angles.

We draw a horizontal line to represent the ecliptic and mark the Sun at S. South of the ecliptic draw a line inclined at 10 times Mars' orbital inclination, at 18·5°. Now Mars itself must lie in the orbital plane MS,

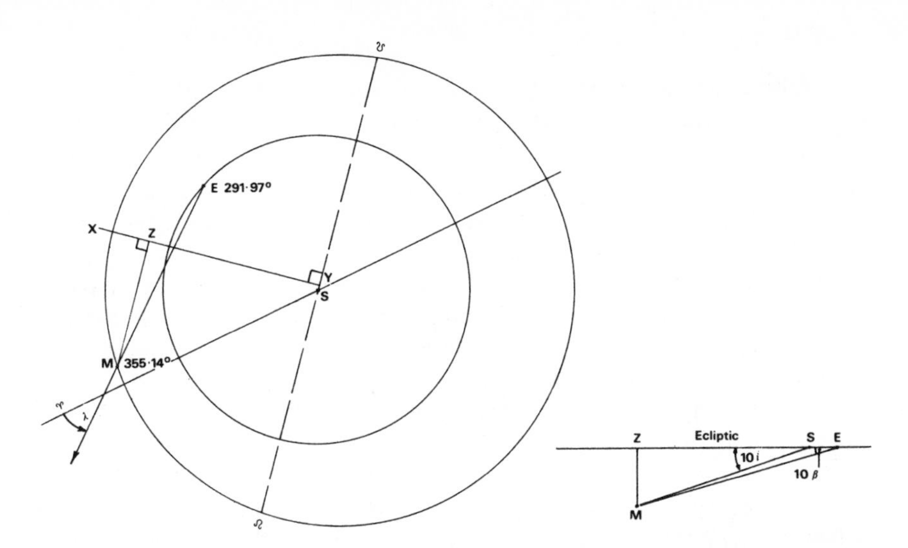

Fig 61 The orbits of Earth and Mars, with the construction lines for calculating Mars' position at 1975, July 15. 0d.

but how far should it be marked from the Sun? If we bisect the line of the nodes (note that it is not quite at the Sun) and mark the position Y, we can draw a line at right angles to this, YX. If Mars were at X it would be furthest south of the ecliptic. As it is, if we drop a perpendicular to XY from Mars at M on the orbital diagram, meeting XY at Z, the distance YZ represents the actual distance Mars has moved along the 'side view' of the ecliptic. So on our side view, drop a perpendicular to cross the inclined orbit SM from Z, where SZ is the same distance on both diagrams. Now, on the side view the line ZM is the actual distance in AU (multiplied ten times) that the planet is south of the ecliptic. By marking the Earth at E on the ecliptic such that the actual distance E to M is the same as on the orbit diagram, we can measure the geocentric latitude 10β on the diagram.

The actual angle on this diagram is $-16{\cdot}2°$ so $\beta = -1{\cdot}62° = -1°\ 37'$.

A mathematical way of finding the latitude can be derived from this diagram, and we can use this if we prefer—it avoids having to multiply the angles by ten with the error this implies. The geocentric latitude is given by:

$$\beta = \frac{ai}{S} \operatorname{Sin}(L - \Omega) \text{ degrees}$$

where a is the mean distance of the planet from the Sun in AU, i is the inclination of the planet's orbit to the ecliptic in degrees, and S is the true distance from the planet to the Earth, measured from the diagram, in AU.

In this example, Mars is 1·3 AU from Earth, so the latitude is:

$\frac{1{\cdot}524 \times 1{\cdot}85}{1{\cdot}3}$ Sin (355·14−49·33), which works out to −1·76°, or −1° 46′.

From Fig 56, we find the RA is 2h 30m, and the Dec 13° 12′, which must be judged against the ephemeris figures of RA 2h 31·3m, and Dec 13° 12′. Mars is in the southern part of Aries.

We do not need to be so precise for many purposes. But say that a rough calculation shows you that the Moon and Mars will be very close together, or that Mars will pass very close to a star. We can use the more exact method to check whether there is any possibility that Mars will be occulated by the Moon, or itself pass in front of the star.

OBSERVING THE EARTH'S ORBIT

Finally we come to a demonstration of the fact that the orbits of the planets are elliptical. Coupled with this experiment, if we know the value of the astronomical unit in miles, we will find a method of measuring the size of the Sun, Moon, the planets, and even the craters on the Moon if we have a small telescope. The value of the astronomical unit is 149,600,000 km (150 million kilometres) or 92,957,209 miles (93 million miles). This is all the data we need for a most satisfying experiment.

Fig 62 shows an object of unknown diameter D at a known distance *S* in kilometres or miles and which subtends a small angle *a*, to the observer at *E*. The diameter *D* can be found from the fact the $\frac{D}{S} = \text{Tan } a$, so that $D = S \text{ Tan } a$. Now the tangent of small angles is almost equal to the angle expressed in radians, so we can say that $D = Sa$. If we have measured *a* in degrees we can convert to radians by multiplying by $\frac{2\pi}{360}$ which is the same as saying that there are 57·3° in 1 radian. If we measure the Sun's angular diameter on October 3, when its distance is about 1 AU we find that it is 0·535°. The Sun's diameter, according to this measurement, is 0·534×150,000,000/57·3=1,394,000 kilometres (the actual value is 1,392,000 km).

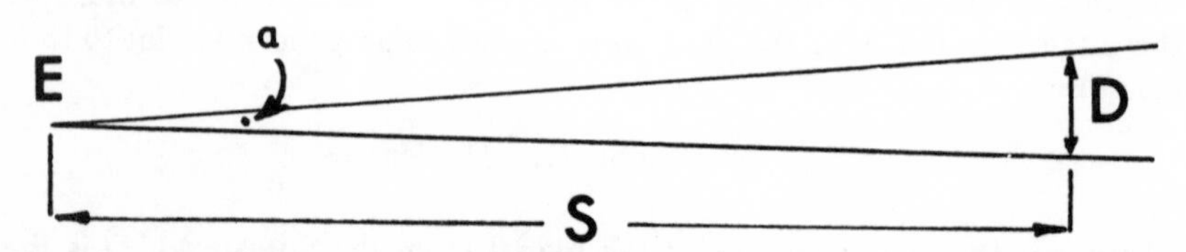

Fig 62 Calculating the diameter D of an object at distance S which subtends an angle of a at E.

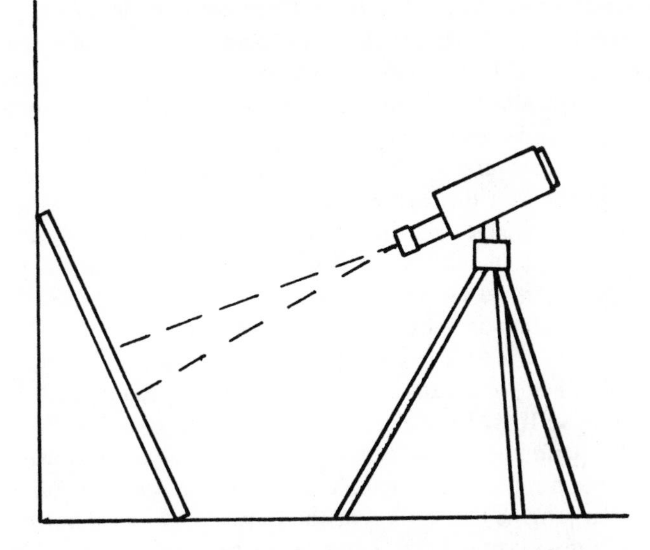

Fig 63 Projection of the Sun's image on to a screen from a pair of binoculars (above), showing the positioning of the screen and (below) the marking, to measure the angular diameter of the Sun's disc by timing its movement across a line on the screen.

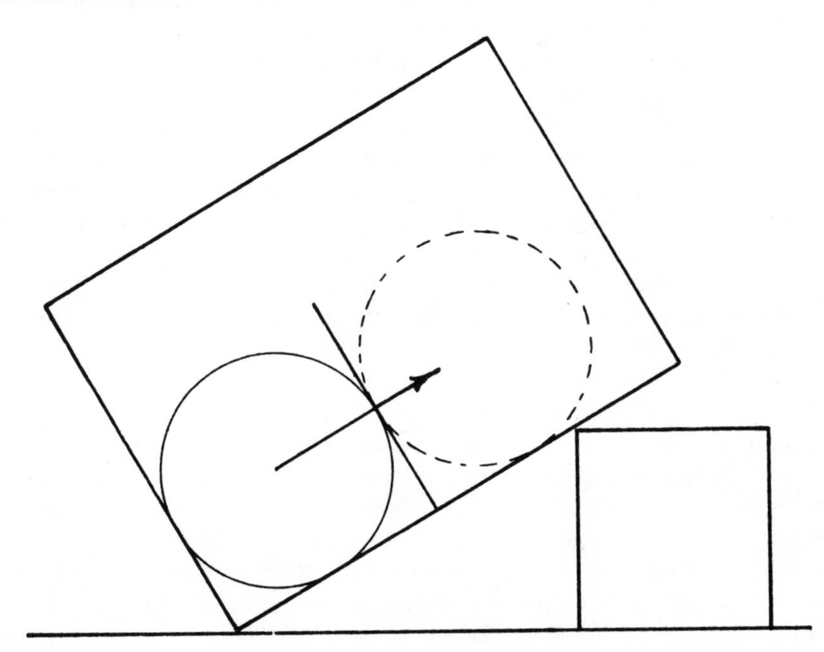

Since the Sun's actual diameter does not change, but the distance does, we can convert the angular diameter of the Sun into the Sun's distance, and plot the Earth's orbit to scale around the Sun. This experiment is carried out at intervals throughout the year—at any of those precious moments when we can see the Sun.

But how do we measure the Sun's angular diameter? Since we must NEVER look at the Sun through binoculars or telescope, we use a projection of the Sun's image to measure its diameter. This can be done without too much trouble as long as we have a pair of binoculars or a simple telescope which we can mount on a tripod.

Mount the binoculars as shown in Fig 63 (left) with one lens covered so that the image of the Sun from the other lens falls on a screen made from a piece of paper pinned to a board. The screen should be shaded as much as possible so that the Sun's image can be clearly seen. Focus the image carefully, with the binoculars or telescope about 1m away from the screen to obtain an image some 20cm to 25cm across. The larger the better, in fact, as long as the outline of the Sun's image does not become too blurred. (You will, of course, plot the position of any sunspots you can see, for comparison a few days later.)

Watch the movement of the Sun's image for a while. If it is afternoon, it will be climbing on the screen, because the image is inverted. Prop one corner of the screen on a box until the Sun's image is climbing up the screen with its lower edge on the bottom of the screen, as shown in Fig 63 (right). You will have constantly to alter the position of the binoculars so an adjustable tripod head is useful. When everything is set, position the Sun's image so that the trailing edge of the disc is not yet quite on the screen. At the instant it is, as shown on the right of Fig 63 with the solid circle, draw a mark on the leading edge. Once more line up the image so that it is about to fit between the trailing edge and the line you have just drawn. You must now time the image's movement from this point to the instant it has moved one diameter, when the trailing edge of the image is on the line, as shown in the diagram with a broken line. Repeat this timing as many times as you can, using a stopwatch, or the most exact timer you have, and take the average of these times. Record the value against the date.

To convert this time to an angle we merely use the fact that the Earth turns at 1° every 4 minutes relative to the Sun. If the Sun were on the equator, therefore, and it took 2 minutes for the image to move one diameter, the Sun's angular diameter would be 0·5°. But the Sun is unlikely to be on the equator, and we wish to do the experiment at various times of the year, so we must apply the correction we met before, and multiply the time by the cosine of the Sun's declination for the date in question. To find the angle in degrees we multiply the time in minutes by Cos Dec, ignoring the + or −, and divide by 4. But if we measure the time in seconds, multiply by Cos Dec, and divide by 4, we shall obtain our answer directly in minutes of arc. You should be particularly careful to obtain results when the Earth is at perihelion and aphelion. When you have collected a years' results you can produce a table of Earth-Sun distances by taking the Sun's angular diameter at a distance of 1 AU as 32·04′ and expressing the other angles as a percent of this value, remembering that the larger the diameter, the nearer the Earth is to the Sun.

We have covered a lot of ground in this chapter, and when you have thoroughly mastered all the techniques described, you will be able to carry out a great many very satisfying experiments of your own. It may even have inspired you to make a more detailed study of the movements of the planets. By patient observation over a few years of the planets at their elongations, oppositions etc, using only the equipment you have made from this book, you could plot their orbits to scale and discover for yourself their shape, and, following the footsteps of the great Kepler, formulate his famous laws for yourself.

EXERCISES ON CHAPTER 8

1 Using the simple method of plotting Mars' orbit, ie using the mean longitudes only, assuming concentric orbits and ignoring the inclination of Mars' orbit, calculate Mars' position in RA and Dec on 1975 July 15·0d and compare with the more precise method used as a worked example in this chapter.
2 At 1981 March 27d 00h two bright starlike objects are seen side by side in the centre of Virgo. What are they?
3 What is the date of the inferior conjunction of Venus in 1975? Did it pass behind the Sun's disc?
4 Find Jupiter's position in RA and Dec for 1975 November 9·0d.
5 Mercury is at superior conjunction at 1975 August 1. Is the following elongation eastern or western? Will Mercury be a morning or evening star then? Will this elongation be a particularly good time to look for Mercury?
6 Assuming Bode's Law holds good for the 10th 11th and 12th as yet undiscovered planets of the solar system (not counting Bode's missing fifth planet where the asteroids are) what are their distances and periods of rotation about the Sun? If planet 12 was at opposition at 1974 July 11d 12h 00m, when is it at opposition in 1975?
7 Herschel discovered Uranus in 1781. Where was it approximately, at the time of his discovery? (Uranus is so distant from Earth the geocentric longitude is almost the same as heliocentric.)
8 A crater on the Moon takes 2·8 seconds to cross a wire positioned north–south in the field of view of a telescope. If the Moon's declination is +28°, and its distance from Earth is 365,450km, what is the distance across the crater?

9
The Double Planet

Ancient astronomers believed the Earth to be the most important and central part of the Universe, whereas we now realise that neither the Earth, nor its primary star, the Sun, nor indeed the galaxy in which that star is situated are particularly important in the cosmological scale of things. In our solar system, however, there is one claim the Earth has to fame from the astronomical point of view: it is a double planet. Its companion, the Moon, is very large in proportion to the size of the primary planet, and in fact the ratio of the sizes and masses of the two bodies is much smaller than any other planet and its largest satellite.

As the Earth and Moon revolve round the Sun, the Moon appears to go round us. A little later when we consider the calculation of the Moon's motion we shall assume that both the Moon and Sun go around the Earth. For our purposes it makes no difference that this is not the true state of affairs.

The Moon revolves about the Earth in a lunar month. There are about 13⅓ lunar months in a year, which is why the *synodic month,* as it is called, is not used for our calendar. As the Moon moves along its orbit from west to east it arrives at positions on the same side of the Earth as the Sun, and, half a synodic month later, on the opposite side of the Earth to the Sun, so that the Moon is in conjunction and opposition at almost fortnightly intervals. During those two weeks the Earth has moved round the Sun by some 14°, so that the starry background to successive full moons, for example, is considerably different. The Moon in fact completes a 360° revolution about the celestial sphere some two days earlier each synodic month, which means that its sidereal period is about 27 days. If you divide 360° by 27 days you will find that in 1 day the Moon moves over 13° across the celestial sphere compared with the Sun's 1°. This is such a rapid change of position that in an hour the Moon moves by more than its angular diameter. On the occasions when it occults, that is, passes in front of a star, its approach to the star is easy to see and is a fascinating sight.

Like the Earth, one half of the Moon is in darkness and the other in sunlight. But from the Earth, the amount of the sunlit surface of the Moon that we can see depends upon its position in the sky relative to the Sun. When the Moon is in conjunction with the Sun, the sunlit face is the opposite hemisphere to that facing the Earth. At new moon, therefore, it is night on the Earth-facing side of the Moon. By the time the Moon has turned through half its path around the Earth, the sunlit face is seen from the Earth. Since the Apollo Moon landings it has become widely known that the Moon presents the same face towards the Earth throughout the month because it rotates on its axis at the same rate as it revolves about the Earth. This means that the Earth is almost fixed in the sky as seen from the Moon, but it does not mean that the far side of the Moon, which has been misleadingly referred to as the 'dark' side, is

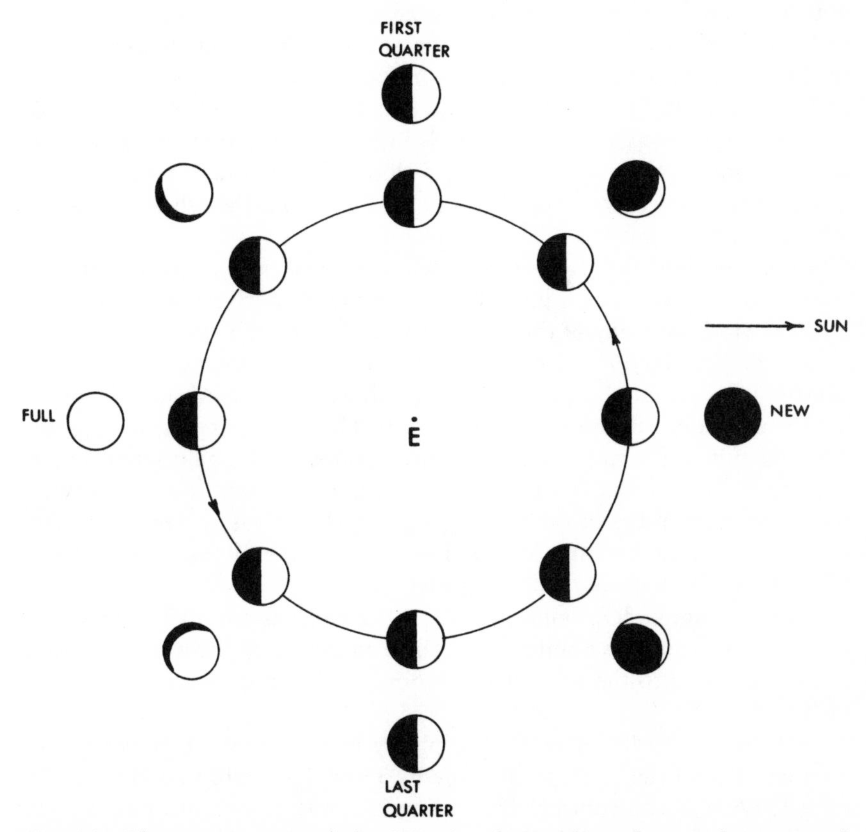

Fig 64 The appearance of the Moon's phases throughout a lunar month (turn the page to place E beneath each position of the Moon in turn).

always in darkness. At new moon, the side turned away from the Earth is fully in the sunlight.

The position is illustrated in Fig 64 which shows the Moon at eight positions around its orbit, together with the appearance of the Moon as seen from the Earth. The characteristic changing shape of the Moon is called its *phase,* and is related to the Moon's age in days since new moon. The phase depends upon the position of the terminator, which is the line across the Moon's disc where the sunlight just reaches. The various phases occur a little later than shown in the diagram due to the Sun's apparent easterly motion around the Earth.

The new moon is always invisible except when it passes across the Sun's disc during a solar eclipse. In common parlance, the Moon is often called 'new' when in fact it is two or three days 'old'. To see the incredibly thin sliver of the Moon's crescent when it is less than one day since new is very difficult, and depends upon favourable conditions, particularly in the inclination of the ecliptic to the horizon at sunset or, if you are looking for the Moon in the last hours before new moon, at sunrise. It is very similar to finding Mercury. The best time of the year for northern observers to look for the Moon when a few hours old is in March. In the morning early in Autumn you may be able to find the thin C of the old Moon before it disappears at new moon. For southern observers, it is the other way round (and the very young Moon appears like a letter C).

An interesting and surprisingly accurate method of calculating the Moon's age for any date is given below, without its derivation. The answer given is within one day of the correct value. To find the age of the Moon: divide the year by 19 and note the remainder (ie reject multiples of 19). Multiply this remainder by 11. Reject multiples of 30. Add the remainder to the number of centuries in the year divided by 3, ignoring the remainder, and add this to the number of centuries divided by 4, again ignoring the remainder. Now add 8. Then subtract the number of centuries in the year. Now add the number of the month, counting March as 1, April as 2, and so on. Now add the number of days in the date. If necessary, reject multiples of 30, and that is the answer!

A quick example: On January 1, 1941, Bremen was heavily raided. (Two days later, Cardiff was bombed.) Bombing raids during the war were often timed according to the state of the Moon. In this case, what was the age of the Moon?

1941 divided by 19 leaves 3. Multiply by 11 and reject 30, giving 3. The number of centuries is 19, so dividing by 3 and 4 gives 10 to add to the 3; total 13. Add 8 and subtract 19, leaving 2. January is month number 11, which gives 13, and add one more for the day, so the Moon was 14 days old, just about full moon.

The Moon's Orbit

The Sun appears to move around the ecliptic at the rate at which the Earth in fact moves round the Sun, just under 1° per day. The Moon travels round the ecliptic at just over 13° per day, so that it is constantly overtaking the Sun at a rate of about 12° per day. The Moon will therefore be at first quarter, that is 90° east of the Sun (see Fig 64) at $7\frac{1}{2}$ days after new moon. At an age of 9 days, for example, it will be $9 \times 12 = 108°$ east of the Sun, and so on. If we do not know the date of new moon, we could choose any date and use the method above to find the Moon's age, multiply the result by 12, and that gives the angle we must add to the Sun's longitude to find the Moon's position, very approximately.

But for many purposes, the error in the Moon's mean position from this calculation which can be up to about 8°, is far too great. If we want to predict an occultation or an eclipse, we must be much more precise. The rest of this chapter is devoted to the more precise calculation of the Moon's motion and will involve quite a lot of simple arithmetic.

The rewards for your efforts are great, however, because you will be able to predict eclipses for many years to come—even finding out from what part of the Earth they can be seen—and with a little patience you will be able to forecast close approaches or occultations of stars or planets by the Moon.

We can start by looking at the Moon's orbit as shown in Fig 65. It is inclined to the Earth's orbit (the ecliptic) at an angle of average value 5° 8′ 43″ (5.15°). For half the lunar month the Moon is south of the ecliptic and for the other half it is north. As it crosses from south to north it passes through the ascending node and then has a positive ecliptic latitude until it passes through the descending node to south of the ecliptic, when it has a negative ecliptic latitude.

The ecliptic longitude of the nodes of the Moon's orbit changes much faster than those of the planets. Whereas we could ignore the very slow change in position of the nodes of the planetary orbits, the Moon's nodes move round the Earth by just over 19° per year. They *regress*, that is, they have retrograde or east to west motion, and this must be remembered carefully during our calculations.

The Moon's orbit is basically elliptical, and has an average eccentricity of 0.05490. The Moon therefore is closer to Earth, and moving faster at one point in its orbit, called *perigee*, than when at the greatest distance from Earth, *apogee*. The corresponding points on the planetary orbits, perihelion and aphelion, are virtually fixed in space, but the line joining the perigee and apogee of the Moon's orbit (the line of the apsides) increases in ecliptic longitude by over 40° each year.

In Fig 65 the ascending node of the Moon's orbit is approaching the

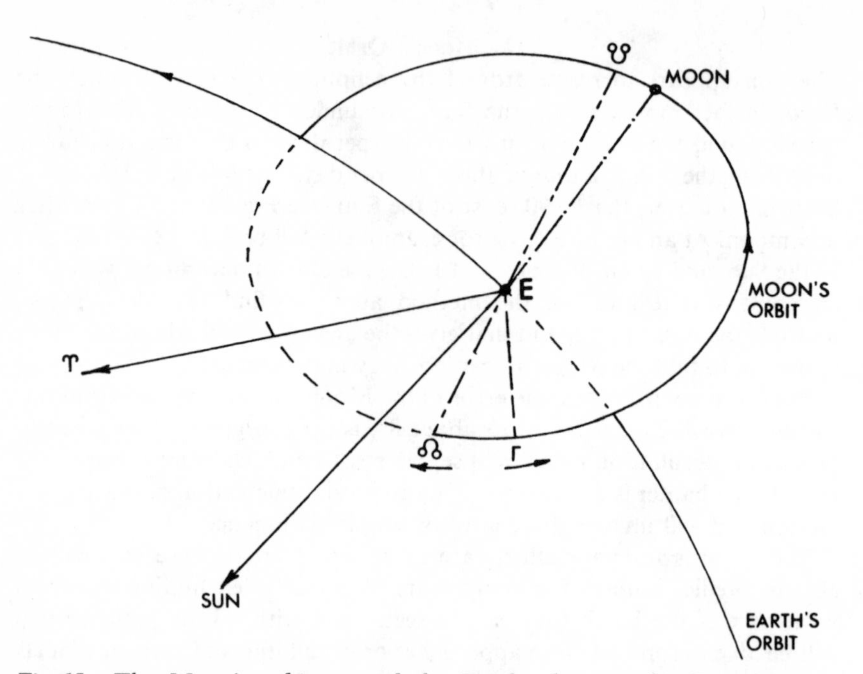

Fig 65 The Moon's orbit around the Earth, showing the First Point of Aries, ♈, the lunar perigee, Γ, the ascending node, ☊, and their relative motions.

same longitude as that of the Sun, and as this is about 45° east of ♈, this is its approximate longitude. The Moon itself has just passed full, and is approaching the descending node, ☋, so it was probably eclipsed by the Earth's shadow. The Moon is also about 90° past perigree (marked with a capital gamma, Γ). The difference between the Moon's longitude and perigree is called the *anomaly* and is measured westwards from the Moon. (The anomaly is also used to refer to the difference of the position of the Sun from the Sun's perigee.)

As in the case of the planets, this difference, arising from the ellipticity of the orbit, means a correction to the mean longitude called the equation of the centre. The correction is much larger than in the case of the planets due to the change in mean eccentricity of the Moon's orbit, so that whereas with the planets we could safely ignore the smaller terms in the series of corrections which must be applied for absolute precision, in the Moon's case, we must apply at least one more term.

The Moon's orbit is disturbed by the Sun, in particular, by an amount which varies according to the position in its orbit. This has the effect of changing the eccentricity, and is known as *evection.*

Other disturbances have relatively significant effects on the Moon's position so that we will have to take them into account. These include the *variation,* a result of the Earth's interference with the influence of the Sun on the Moon, and the *annual inequality* due to the change in distance from Moon to Sun over the Earth's yearly motion.

We will ignore the thousand or so other small corrections which should be made, since taken all together they add up to only a few seconds of arc. In working through the position calculations it is wise to use at least three places of decimals when adding figures, and possibly four when multiplying. The third place of decimals will be quite useful to round off the second one, and the accuracy of our methods will just about justify this.

Calculating the Moon's position

For the purposes of this calculation, we can assume that the Sun and Moon move round the Earth, so we will not need scale drawings of the orbits as we used for the planets, and the distances of the Sun and Moon are only taken into account to determine the apparent sizes of their discs.

The procedure is described in a sequence of numbered steps. At the end of this sequence you will have calculated the Moon's ecliptic longitude and latitude, and this can be translated into RA and Dec from Fig 56, but the values will be the position as seen from the centre of the Earth. Later we will see how we can correct these geocentric values to account for our position on the Earth's surface.

1 Calculate the Earth's mean heliocentric longitude, starting from the 1971 January 0d 00h position (99·011°), then correct for the equation of the centre. (NB. Do not use the 1971 Jan 0·5 position as for the planetary calculations.) Add 180° (or subtract) to give the Sun's true geocentric longitude, λ_s.
2 Starting from the 1971 Jan 0d 00h position (314·160°) find the Moon's mean geocentric longitude λ_m, by adding the mean yearly, daily and hourly motions.
3 Similarly, find the mean longitudes of perigee, Γ, and ascending node, ☊.
4 The mean anomaly of the Moon is given by $A = \lambda_m - \Gamma$. The mean elongation is given by $D = \lambda_m - \lambda_s$. (A and D are measured westwards.)
5 Add to λ_m,A, and D the correction for evection: 1·3 Sin (2D−A).
6 Add to λ'_m, A′, and D′ (the first corrected values), the correction for the annual inequality: −0·2 Sin A_s, where A_s is the Sun's anomaly $(\lambda_s - 282{\cdot}442)°$.

7 Add to the second corrected values, λ''_m, A″, and D″, the correction for the equation of the centre: 6·3 Sin A″+0·2 Sin 2A″.
8 Add to the corrected values, λ'''_m, A‴, and D‴, the correction for the variation: 0·7 Sin 2D‴. The corrected longitude is λ_M, the Moon's true geocentric longitude.
9 The Moon's geocentric latitude is given by: $\beta_m = 5{\cdot}15 \text{ Sin } (\lambda_M - \Omega)$.
10 To correct for the inequality of latitude, add the following angle which is given in minutes of arc, not degrees: 10 Sin (2D‴$-\lambda_M+\Omega$).

Note that throughout, until the last step—the latitude correction—we are working in degrees.

There are plenty of chances for mistakes in so many steps: be careful with the signs of trig functions (see appendix); do not forget to include the leap days when adding up the total time since 1971 Jan 00d 00h; and don't forget to subtract the retrograde motions (ie with the Earth's net mean annual motion, and the nodes).

To help a little with the Moon's position over the year, the net change in longitude for 50, 100, and 200 days and one year of 365 days, has been given in the appendix. These values can be added as required to close to the date in question.

Finally, tabulate your working; you could copy the layout used in the worked example which follows.

Calculate the Moon's position at 1975 June 16d 12h.

	Years	Days	Hours
Total elapsed time since 1971 January 00d 00h:			
Years since 1971:	4		
Days since 1975 Jan 0d to June 1 (from table)		152	
Remaining days to June 16		15	
Leap year days since 1971		1	
Hours since 00h			12
Total elapsed time:	4 years	168 days	12 hours

Earth's and Sun's positions:

Earth: mean longitude at 1971 January 0d 00h:	99·011°
Corrections: for four years' motion, add 4×−0·239	−0·956
for 168 days, add 168×0·9856	165·581
for 12 hours, add 12×0·0411	0·493
Earth: mean longitude 1975 June 16d 12h	264·129°

Correction to mean position for equation of centre:

$$1{\cdot}92 \text{ Sin } (264{\cdot}129-102{\cdot}441) = +0{\cdot}603°$$

Add correction to mean longitude; Earth's true longitude: 264·732°

Sun's true longitude, $\lambda_s = 264{\cdot}732 - 180 = 84{\cdot}732°$

Sun's anomaly, $A_s = 84{\cdot}732 - 282{\cdot}442 = -197{\cdot}71° = 162{\cdot}29°$

Moon's mean longitude:

Mean longitude at 1971 January 0d 00h	314·160°
Corrections: for four years add 4 × 129·385	517·540
for 100 days (from table)	237·640
for 50 days (from table)	298·820
remaining days, add 18 × 13·1764	237·175
for 12 hours add 12 × 0·549	6·588
Total:	1611·923
Reject multiples of 360°	−1440·000
Mean longitude 1975 June 16d 12h:	171·923°

Mean longitudes of perigee and node:

	Γ		☊
Mean position, 1971 Jan 0d 00h:	343·199		326·009
Four years' motion, add: (4 × 40·662)	162·648	(4 × −19·328)	−77·312
168·5 days' motion: (168·5 × 0·1114)	18·771	(168·5 × −0·05295)	−8·922
Totals:	524·618		239·775
Reject multiples of 360°	360·000		
Mean positions 1975 June 16d 12h	164·618°		239·775°

Anomaly and elongation:

Moon's mean anomaly, A; 171·923 − 164·618 = 7·305°

Moon's mean elongation, D; 171·923 − 84·732 = 87·191°

Corrections to Moon's mean longitude, anomaly and elongation:

	λ	A	D
Mean values:	171·923	7·305	87·191
Evection: 1·3 Sin (2 × 87·191 − 7·305) = +0·291	172·214	7·596	87·482
Annual inequality: −0·2 Sin 162·29 = −0·061	172·153	7·535	87·421
Equn. of centre: 6·3 Sin 7·535 +0·2 Sin (2 × 7·535) = +0·883	173·036	8·418	88·304
Variation: 0·7 Sin (2 × 88·304) = +0·041	173·077	8·445	88·331

Moon's true geocentric longitude, 1975 June 16d 12h = 173·077° = 173° 5′

Moon's geocentric latitude:

$\beta = 5{\cdot}15 \,\text{Sin}\,(173{\cdot}077 - 239{\cdot}775) = -4{\cdot}729° = -4°\,43{\cdot}7'$

Latitude inequality: 10 Sin $(2 \times 88{\cdot}331 - 173{\cdot}077 + 239{\cdot}775) = -8{\cdot}9$ arc min.

Moon's true geocentric latitude, 1975 June 16d 12h = $-4°\;52{\cdot}6'$

The position given in the ephemeris for this particular moment was longitude 173° 05′ 04·17″, latitude −4° 50′ 50·12″. We cannot expect that close an answer each time we carry out the calculation, particularly in longitude, but we should always be within about 20 minutes of arc.

We can convert the ecliptic co-ordinates into RA and Dec from Fig 56 if we wish and we can find the Moon's position in a star atlas. On this occasion we find the Moon just north of the 5 magnitude star 87 Leonis, and as we will see later on, it is close enough to expect that it will occult this star as seen from northern latitudes of Earth.

The calculation also shows that the elongation is just under 90° west of the Sun when it will be at the first quarter. The Moon overtakes the Sun at the difference in their mean hourly rates, which comes to 0·508° per hour, or 0·508′ per minute, so the time of the first quarter will be the remaining 1·67° divided by 0·508° per hour = 3·29 hours later, at 15h 17m. This is, in fact 19 minutes late due to factors we have ignored, but it is quite satisfactory.

A lunar position calculator

Just as we used the basic method of calculating the positions of the planets to make a position calculator, we can adapt the method we use to find the Moon's position to make a calculating device. If this is made carefully it is capable of surprising accuracy, and will give you all the results you need except for the most accurate work. It will enable you to find the dates, and even the approximate time, of eclipses and for this function alone it is well worth making. To work out the circumstances of an eclipse without a lot of fruitless work means that you must know to within a day or two when the eclipse will take place. The calculator will enable you to do this for eclipses long into the future.

Like the planetary calculator, this one consists of a number of discs on a common centre, with a clear plastic cursor. But this time, as in our arithmetical method, the distances of the Sun and Moon are not important, so we can make the discs the size we find most convenient to use.

The calculator is illustrated in Fig 66. On the left of this figure, the basic layout is shown. There are 6 discs, cut out of thin card. The smallest diameter of the inner disc should be 125mm, and each disc should be at least 25mm diameter larger than the one on top, so that the outer disc will be at least 250mm. There is a lot of marking to do on most of the discs,

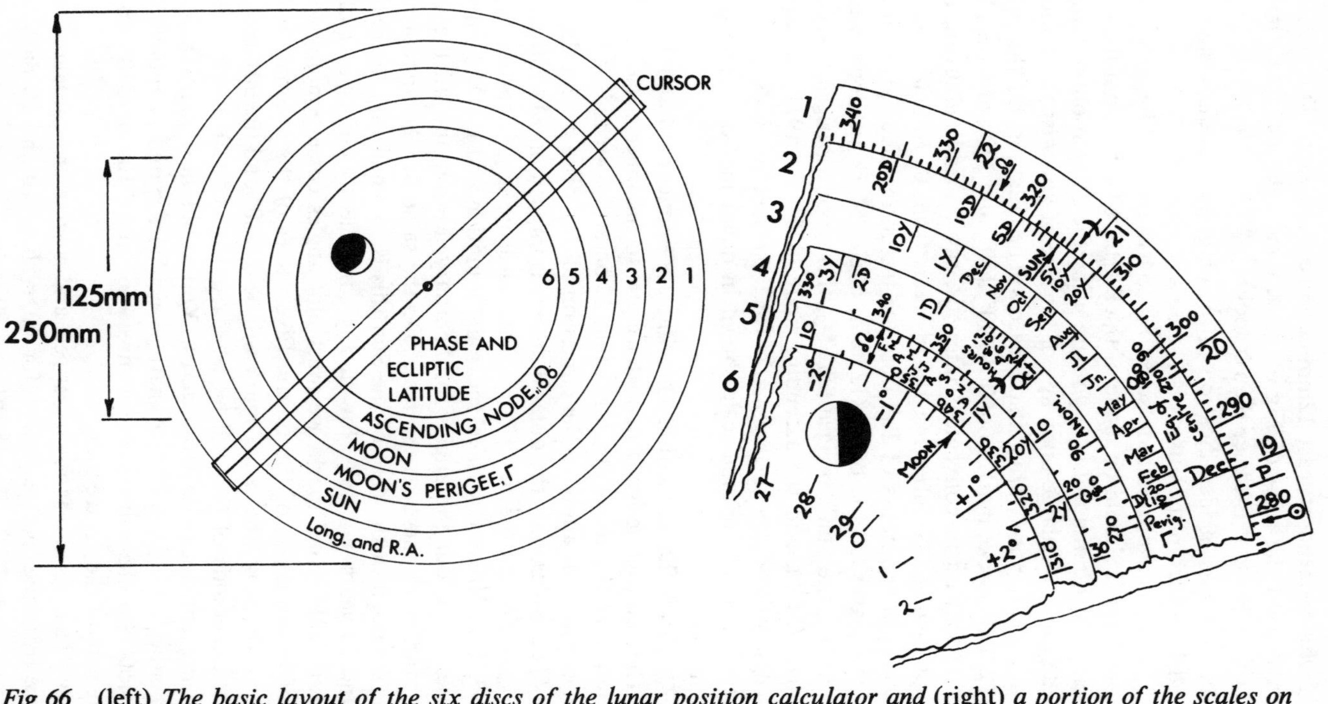

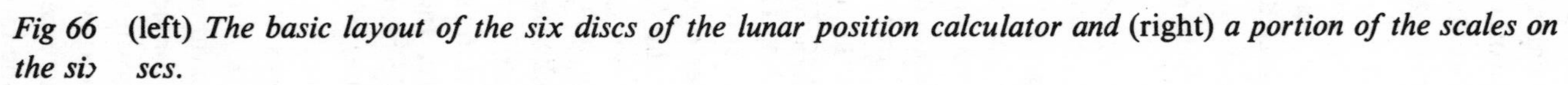

Fig 66 (left) *The basic layout of the six discs of the lunar position calculator and* (right) *a portion of the scales on the si› scs.*

which is why we need at least 12mm clear between each. The section of the calculator on the right of Fig 66 gives some idea, but don't let this put you off, because it is easier than it looks!

For the purposes of identification in the diagram, the discs have been numbered 1 to 6 from the outer disc.

DISC 1 This is the longitude scale. Since it forms the backing to the whole calculator it should be made of thick artist's card. From the centre draw a circle of 125mm radius if you are using the 250mm outer diameter suggested (you can enlarge upon any of the sizes if you wish) and another of the radius of disc no 2, which in this case will be 112·5mm. On the outside of the 112·5mm circle, mark from 0° to 90°, 180°, and 270°, and then back to 0° anticlockwise in convenient intervals, say 5°, and subdivide these into single degrees. Number the 10° intervals from 0° to 350°. Mark in the positions of the hours of RA opposite the corresponding longitudes, making the marks on the outer (125mm) circle. Mark the longitudes of the Sun (279°), the Sun's perigee (282·5°), the Moon's perigee (343°), the Moon itself (314°), and the ascending node (326°) for 1971 January 0d 00hrs. These will be the starting marks for the respective discs. In Fig 66, the portion of the longitude scale shown in detail includes some of the starting marks.

DISC 2 This is the Sun's disc. This time use thinner card and draw two circles. The outer circle has a diameter equal to the inner circle on disc 1 (125·5mm), and the inner circle is again 12·5mm less in radius than the outer one. This will apply to the circles on all the discs except the inner one, no 6, on which only the outer diameter is important. Mark the Sun's symbol on a line on the outer circle, then measure angles corresponding to 1, 2, 3, 4, 5, 10, and 20 days' mean daily motion and mark them anticlockwise. Work out the angles for February, March and all the other months, and mark these on the scale. Do the same for the annual mean motion (although in the case of the Sun this is so small that intervals of 5, 10, and 20 years will do). Now at any conveniently empty part of the scale near to the 'Sun' mark, we must plot angles corresponding to the values of the equation of the centre. Five values of the anomaly will be enough for the Sun. The centre mark will be 0°/180° and the 90° and 270° marks at either side of this will be 1·92° from the 0°/180° mark (the 90° mark being anticlockwise from zero). You could also mark half this interval either side of the 0°/180° mark, which correspond to anomalies of 30° or 150° (in the anticlockwise direction) and 210° or 330° on the opposite side of the centre mark.

DISC 3 This is the Moon's perigee and angles are measured anticlockwise from the 'perigee' mark for intervals of 10 and 20 days, each of the months, and 1, 2, 3, 4, 5, 10, and 20 years.

DISC 4 This is the Moon itself. The procedure for marking this disc is the same as for the others, but this time you must mark 2-hour intervals as well. Up to 12 hours will suffice, because you can always add two lots of hours when using the calculator. This scale should be marked anti-clockwise on the outer circle of the disc. You will find a convenient gap in the scale just below the 'October' mark to put the anomaly scale. Mark a zero and measure 6·3°, 5·5°, and 3·2° either side of it. These seven points should be labelled 270°, 240/300°, 210/330°, 0/180°, 30/150°, 60/120°, and 90° in an anticlockwise direction. A second scale, simply 0° to 360° at 5° intervals, should be marked in a *clockwise* direction on the inner circle, starting from 0° at the Moon mark.

DISC 5 The node's disc must be marked *clockwise* for the angle for each of the months (you can ignore the days) and years. On the inside circle mark a 0 to 360° scale at 5° intervals, this time anticlockwise, from an arrow at any convenient point on the circle.

DISC 6 This is the latitude scale, and gives the solution to the equation $\sin \theta = x/5{\cdot}15$, where x is 1, 2, 3, 4, and 5·15°, various possible latitudes above and below the ecliptic. From a zero marked 'Moon', mark angles of 22·8°, 35·5°, 51°, 76°, and 90°, in both directions and mark the same angles either side of an unlabelled zero 180° from the 'Moon' zero. Label these angles (reading anticlockwise from 'Moon') 1, 2, 3, 4, 5·15° (at the 90° mark), 4, 3, 2, 1, and zero (at the 180° mark) continuing −1, −2, −3, −4, −5·15, −4, −3, −2, −1° and so back to 'Moon', which is also marked 0°.

About 25mm inside the outer radius of disc 6, draw another circle, and put a 0 mark beside the 'Moon' mark on this disc. Now multiply the Moon's mean daily increase in elongation from the Sun, 12·1908°, by 2, 3, 4, and so on up to 29, to obtain the mean elongation of the Moon throughout the month. The 29 day mark will be just above the 0 day mark, by about one quarter of a day. Mark these angles clockwise around the circle, and number them 0 to 29 days.

The last feature of the calculator is a novelty which does not take part in the solution of problems, but it is interesting, and can be used to estimate the age of the Moon from its appearance. Draw a small circle near the outer edge of disc 6. The diameter of this circle should be as large as can be managed so that it will fit into the 12·2° between imaginary radii from the centre of the disc through each of the daily position marks just described. We will return to this feature in a moment.

Cut a strip of plastic of the same length as the outer diameter of disc 1, with a fine straight line scribed along its centre line. Make the mark clear by drawing along it with indian ink or ball point pen, cleaning off any surplus. Push a drawing pin through the line half way along the strip.

Carefully cut round the outside diameters of the discs, and cut out the little circle near the edge of disc 6. Having pushed the drawing pin through

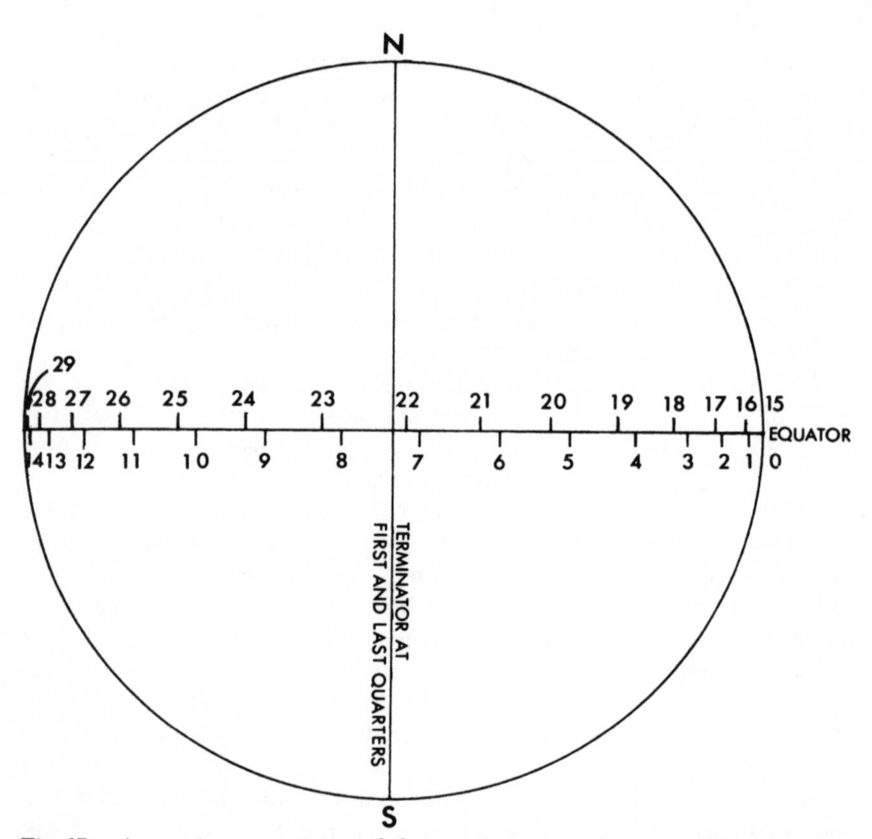

Fig 67 Approximate position of the terminator on the Moon's equator for each day of the lunar month. The terminator must also pass approximately through the poles.

the centre of the plastic pointer, push it through the centre of disc 6, then no 5, no 4 and so on and finally through the thick card no 1 disc. If necessary, push the end of the pin into a small piece of wood at the back of the calculator. Now set the 0-day mark on the inner scale of disc 6 so that it is opposite the arrow marked at 0° on the inner scale of disc 5. The Moon at 0 days, a new moon, is invisible so fill in the small circle with black ink or ball-point pen. Now turn disc 6 until 1 day is opposite the node mark. In the small circle, draw the thin backwards 'C' of the 1-day-old Moon (if you are a northern hemisphere observer, or upside down for the southern hemisphere). Continue to draw the phases for each of the ages of the Moon in days. The approximate position of the terminator for each day of the lunar month is shown in Fig 67, which shows where the terminator crosses the Moon's equator.

This calculator is designed to be used from the outside disc inwards, so that you do not move those which have already been set.

Let us work through the settings for 1982 March 18, at 16h. Put the Sun on disc 2 against its start mark on disc 1. Now move the cursor to '10 years' on the Sun's disc, and then hold it at that spot while you put the Sun under the cursor line, and continue as we did with the planetary calculator. We must add 3 days' motion for the leap years since 1971. To correct for the equation of the centre, estimate the angle that the Sun has passed through since perigee (or measure it on the longitude scale). In this case it is 73·5°. Now put the cursor line on that angle on the equation of the centre scale (you must estimate the position, but on this small scale it is far from critical). Now move the zero mark of this scale (not the 'Sun' mark itself) under the cursor, which moves the Sun forward a little in this case, giving its position as 358·2°.

Set the lunar perigee opposite its starting mark and successively move it in steps to 10 years, 1 year, 3 leap days, March, and finally interpolate the position of 18 days. This gives the perigee's final position of 78·8°.

Holding the first two discs firmly in position, set the 'Moon' at its start. Move it to the positions of 10 years, 1 year, 3 leap days (most important), March, 10 days, 5 days and 3 days, 12 hours and 4 hours. All that brings the Moon to 281·2°. Now if we set the cursor line over the lunar perigee mark on disc 3, we can read the mean anomaly on the inner (360°) scale on the Moon's disc. The anomaly is 218°. Now set the cursor to the estimated position of 218° on the Moon's anomaly (equation of the centre correction) scale, and move the disc back a little to bring the 0/180° mark under the cursor line. Be careful to set the zero of the 'equation of the centre' scale under the cursor, and not the 'Moon' itself. After this correction, the Moon's longitude is 277·4°.

The node is next. Go through the same procedure, estimating the position of March 18 between the marks for March and April. The longitude of the node is 108°. Set the Moon line on the inner disc to point at the 'Moon' itself on disc 4 (using the cursor line) and opposite the node mark read the latitude, +1°. So, on March 18, at 16h, in 1982, the Moon is at longitude 277·4°, latitude +1°.

Finally, set the cursor on the 'Sun', and read the angle with the 'Moon' on the inner scale on disc 4: 282°. Set the 'Moon' line on disc 6 to 282° on the inner scale (the 360° scale) on the node's disc, and opposite the arrow we find that the Moon is just over 23 days old. If we move the disc a fraction to read exactly 23 days, the appearance of the Moon is shown in the little window in the inner disc.

What have we found out from this example? On this date in 1982, just before the Spring equinox (note the position of the Sun) the Moon is approaching its descending node. It is at about RA 18h 30m, so it will be due south at about sidereal time 18h 30m, which at this time of the year is almost exactly 12 hours ahead of UT, so it is south at 6h 30m in the morning from longitude 0° (although this ignores 14½ hours further movement of the Moon, about 8°, making its RA 19h, so it will be 7h 00m on March 19 when it is due south). Its declination will be about −22·5° so it will be very low at Greenwich. At 7 o'clock in the morning at the equinox the Sun will have risen an hour ago, so it will be broad daylight and the Moon's waning crescent will be difficult to see. The Moon has also just passed apogee, so it is almost at its greatest distance from the Earth.

Horizontal Parallax

Because the Moon is in an eccentric and elliptical orbit, its distance from the Earth varies from the mean value of 385,000km (about 240,000 miles), with a maximum at apogee of 406,000km and a minimum at perigee of 357,000km. This means that the apparent size of the Moon as seen from the Earth varies. To the Apollo astronauts, the apparent size of the Earth would also have seemed to be changing, if they had stayed on the Moon for several weeks. Astronomers use the apparent size of the Earth as seen from the Moon to express the Earth-Moon distance. It is given as the angle the semidiameter (the radius) of the Earth's equator subtends at the Moon. It is known as the *horizontal parallax*, and its value is about 1° (some four times the semidiameter of the Moon as we see it).

In Fig 68, the Moon is shown at a distance m from the Earth. The Earth's radius is R, which subtends the angle HP, the horizontal parallax at distance m. From this diagram we can see that $R/m = \text{Sin HP}$. Since HP is very small, if we express it in radians the angle will be approximately equal to its sine, so $R/m = \text{HP}$. Knowing R, and the factor for

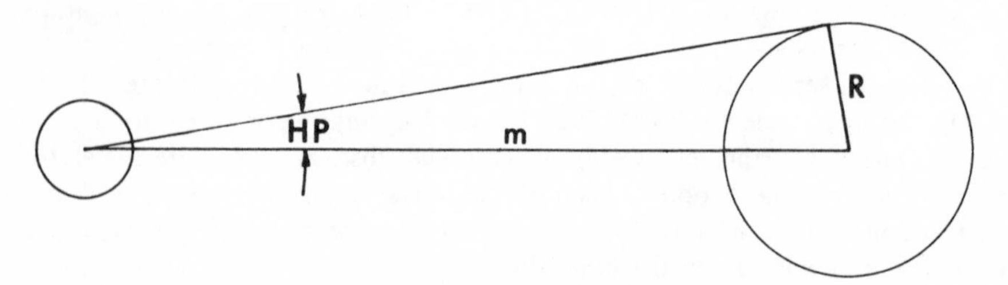

Fig 68 Construction to find the expression for horizontal parallax.

converting minutes of arc to radians we arrive at an expression for the distance from the Earth's centre to the Moon's centre of:

$$m = \frac{21{,}926{,}700}{\text{HP}} \text{ km, where HP is in arcmin.}$$

The mean value of HP is 57·04′, and its precise value at any value of the Moon's anomaly A, is given by:

$$\text{HP} = 57{\cdot}04 + 3{\cdot}11 \text{ Cos A arcmin.}$$

The value can in fact vary due to evection, but we can ignore this.

When we plotted the position of the Sun and planets in our skies we ignored any parallax effect; the planets are so far away that it amounts to only a few seconds of arc at the most. In the Moon's case we cannot ignore parallax. Viewed from one pole, the Moon can be almost 2° different from its position as seen from the other. So if we wish to use our calculations of the Moon's geocentric position to find out its position in the sky as we see it, which is essential if we wish to predict eclipses of the Sun or occulations of stars and planets, we must apply corrections to both ecliptic longitude and latitude to allow for our own longitude and latitude on Earth.

To avoid the confusion of the terrestial and celestial longitudes and latitudes, it is advisable to use the Greek letter symbols for these co-ordinates. Ecliptic longitude is λ (lambda), and ecliptic latitude is β (beta), while on Earth, the observer's longitude is θ (theta) and his latitude is ϕ (phi).

CORRECTION OF GEOCENTRIC CO-ORDINATES

Fig 69 shows an observer at O, at latitude ϕ. The Moon is due south at some place on Earth to the observer's west, so it is west of the observer's meridian by H, the hour angle. Clearly, this will tend to shift the Moon a little to the west of the position as seen from the centre of the Earth C or from the longitude θ at which the Moon is on the meridian. Similarly, because of the observer's position north compared with the centre of the Earth, the Moon will appear to be displaced to the south from its geocentric latitude. The latter is shown on the diagram; β_m is the geocentric value, and β_m' is the latitude as observed at O.

Using the co-ordinates described on this diagram, an approximate value of the corrections to be applied to the geocentric values can be worked out. If we call all co-ordinates south of the terrestrial and celestial equator or the ecliptic negative, to correct the geocentric latitude the following angle must be subtracted:

$$\text{HP (Sin } \phi \text{ Cos } \delta - \text{Sin } \delta \text{ Cos } \phi \text{ Cos H) arcmin.}$$

Where δ is the declination of the ecliptic at the Moon's longitude. If the expression turns out to have negative value, since it is to be subtracted from β, it will, in effect be added.

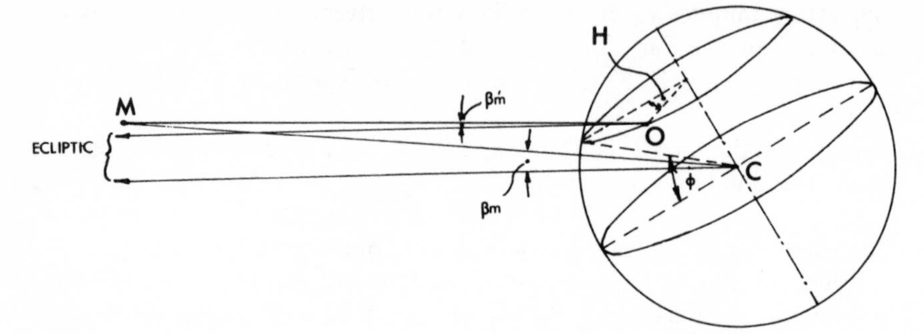

Fig 69 Construction to show why the ecliptic latitude calculated from the values at the centre of the Earth, must be corrected for the observer's latitude on the Earth.

Similarly, the correction which must be subtracted from the geocentric longitude to give the Moon's apparent longitude is:

$$\frac{\text{HP Cos } \phi \text{ Sin H arcmin.}}{\text{Cos } \delta}$$

These expressions are not strictly valid for ecliptic co-ordinates, and to be more precise we should translate the Moon's longitude and latitude into exact values of RA and Dec, but this is another lengthy process, and the gain in accuracy does not warrant the effort in our case. The work may appear involved enough as it is, but if you have worked out the geocentric position even only approximately on the calculator, you will be close enough for many purposes. However, if you are able to predict an occultation or the circumstances of an eclipse as seen from your own garden, and then you actually see it happen, the satisfaction this will bring will make it seem well worthwhile.

AN EXAMPLE

Assume that a calculation of the Moon's geocentric position has given the following data: Moon's ecliptic longitude 70°, latitude + 5°, local sidereal time 7h 40m, Moon's anomaly 162°. What is the apparent position from latitude 60° north?

Plotting the geocentric position on the ecliptic on our chart we find the Moon's RA is 4h 35m and Dec + 22°. The hour angle is therefore 7h 40m − 4h 35m = 3h 5m, which is 46°.

HP is 57·04 + 3·11 Cos 162, which comes to 54·08′. The correction to the geocentric latitude will be: 54·08 (Sin 60 Cos 22 − Sin 22 Cos 60

Cos 46), which works out at + 36·34′. This must be subtracted from the geocentric latitude to give the apparent value + 4° 23·7′.

The longitude correction is given by:

$$\frac{54{\cdot}08 \text{ Cos } 60 \text{ Sin } 46}{\text{Cos } 22}$$

which is 20·98′. This gives an apparent longitude of 69° 39′.

Eclipses

By simple geometry, we can show that the length of a planet's shadow in space is given by L = rD/R−r, where L is the length of the shadow, R is the radius (semidiameter) of the Sun, r is the radius of the planet, and D is the distance from the Sun to the planet. If we work out the value for the Moon, using the mean distance to the Sun, we find that its shadow is about 380,000km long, which quite by coincidence is about the same as the mean distance from the Moon to Earth's surface. This means that the tip of the Moon's shadow, which is in the form of a cone due to the size of the Sun, can just reach the Earth's surface. If we observe the Sun from a point on Earth in that shadow, the Sun is totally eclipsed. If we are not too far away from the point on Earth where the Sun is totally eclipsed, we will see part of the Moon's limb obscuring part of the Sun, and there will be a partial eclipse.

The coincidence that the Sun and Moon appear to be about the same size, despite their completely differing actual sizes and distances, must be the most fortunate oddity of the solar system as far as astronomers are concerned. As a result, when our satellite passes in front of the Sun's disc, it obscures the normally blinding light of the Sun's surface, and reveals the fantastically beautiful glowing outer atmosphere of the Sun, the corona.

Since the Moon is passing between the Sun and Earth at a solar eclipse, it must be the instant of the new moon. A solar eclipse does not occur at every new moon by any means however, because, due to the inclination of its orbit, the Moon can be up to 5° north or south of the Sun. But if the Moon happens to be close to the ecliptic, that is near one of its nodes, there will be an eclipse somewhere on Earth.

The situation is illustrated in Fig 70, which is a view of our double planet as seen from the plane of the ecliptic; but the diagram has to be hopelessly out of scale. The Moon is not quite at a node, but the shadow still falls upon the Earth. At certain critical distances of the Moon from the Earth and of the Moon from the Sun, the shadow cone points at the Earth but is too short to reach it. In such circumstances, observers on

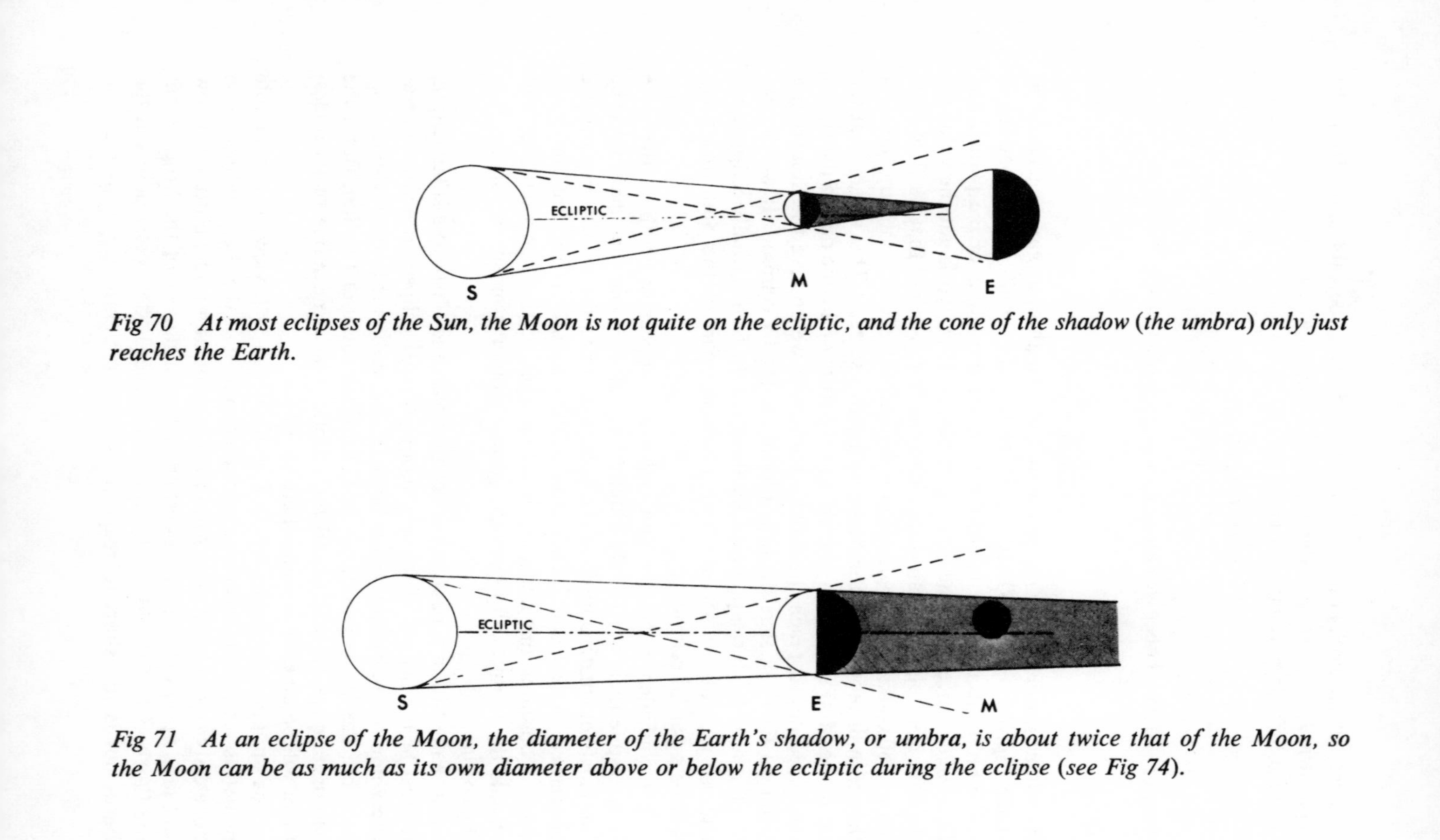

Fig 70 At most eclipses of the Sun, the Moon is not quite on the ecliptic, and the cone of the shadow (the umbra) only just reaches the Earth.

Fig 71 At an eclipse of the Moon, the diameter of the Earth's shadow, or umbra, is about twice that of the Moon, so the Moon can be as much as its own diameter above or below the ecliptic during the eclipse (see Fig 74).

Earth see the black disc of the Moon cross the Sun, but the Moon's diameter appears a little less than that of the Sun, so that the Moon's disc is surrounded by a bright ring of the Sun's disc: this is an annular eclipse. Where the shadow of the Moon is long enough to reach the Earth only at its closest point on the surface, that is, where the Moon appears to be due south, a solar eclipse can be total at this place, but at other places on Earth it will be annular.

In Fig 71 the circumstances for an eclipse of the Moon are shown. Again, the Moon is near its node, but not exactly on it. In the case of a lunar eclipse, the Moon is passing through the cone of the Earth's shadow, and the event can be seen from anywhere on Earth where the Moon is above the horizon. The Moon can also enter the region of space where the Sun is partially obscured by the Earth. In this partial shadow, called the *penumbra*, the Moon appears slightly dimmed. In the total shadow, called the *umbra,* the Moon often becomes a deep red as it is lit by the light refracted through the Earth's atmosphere from the Sun. From the Moon, the eclipse of the Sun must be a splendid sight as the Earth becomes surrounded by a fiery ring of red light. The brightness of this red light during a lunar eclipse can vary considerably due to the state of the Earth's atmosphere: on one famous occasion, a lunar eclipse occurred when there were terrible forest fires burning in Canada. This made the Earth's atmosphere so murky that the Moon completely disappeared during the eclipse.

PREDICTING ECLIPSES

Predicting eclipses is not as difficult as you may imagine. In fact we have already covered all the calculation necessary to do this exciting work. The eclipses of the Sun are the most difficult because, as we have seen, the appearance of a solar eclipse depends upon where you are on Earth.

If we plot the ecliptic longitudes of the two nodes of the Moon's orbit for a year, using the annual motion of the nodes to find the starting and finishing values in any given year, and then plot the ecliptic longitude of the Sun during the same year, as in Fig 72, we can see when the eclipses are likely to occur. If we add to the diagram the lunar months through the year, we can be even more precise about when eclipses of the Sun and Moon will happen.

In Fig 72, the Sun is at the same longitude as one of the Moon's nodes in the middle of June. Just before this happens the full moon is due, so there will probably be an eclipse of the Moon at that time. Towards the end of November, the Sun is once again at one of the Moon's nodes, and the new moon is at about this time, so there will be an eclipse of the Sun. An eclipse of the Sun is also likely to occur at the June conjunction with

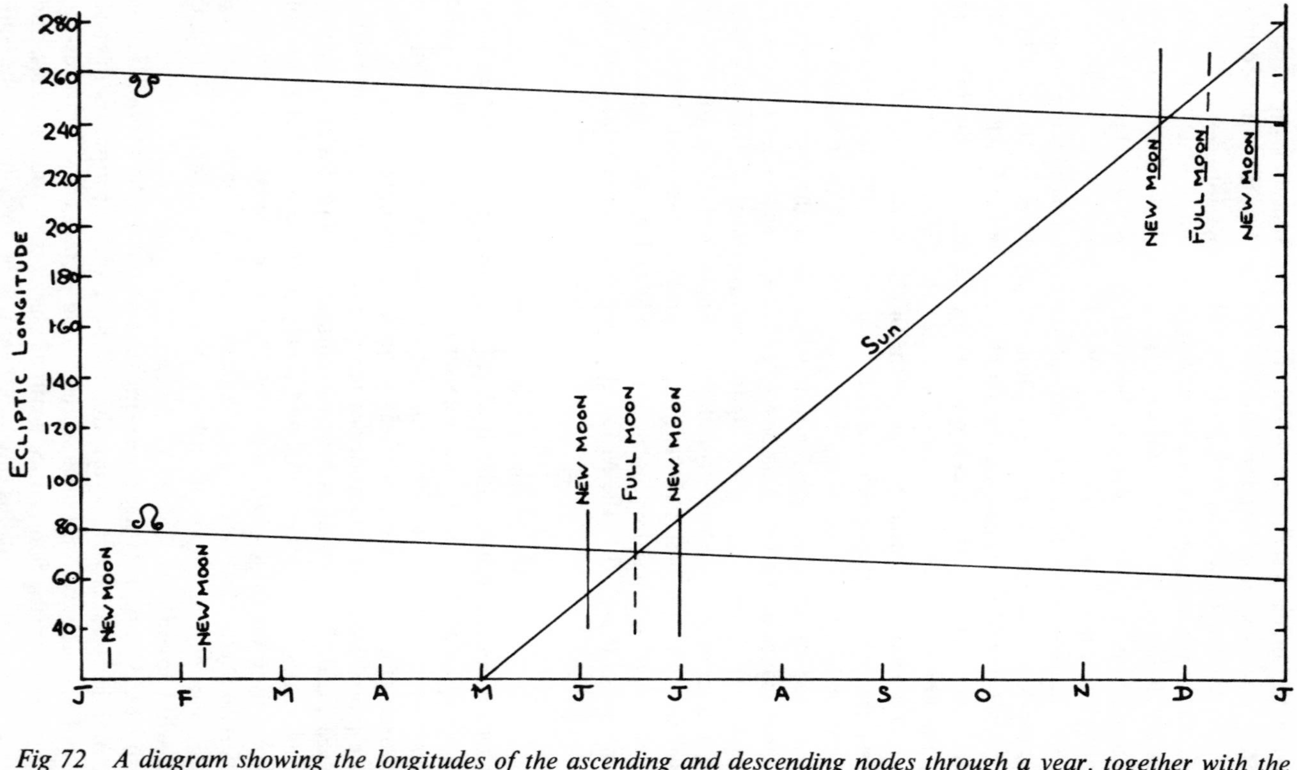

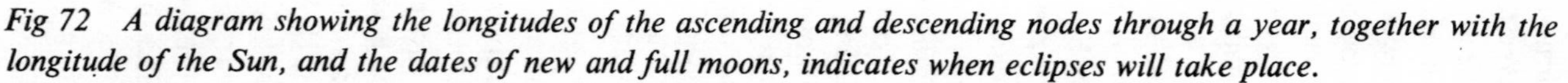

Fig 72 A diagram showing the longitudes of the ascending and descending nodes through a year, together with the longitude of the Sun, and the dates of new and full moons, indicates when eclipses will take place.

the node, probably towards the end of the month.

We can also use the Moon position calculator to predict eclipses of the Sun and Moon with considerable accuracy. Set the Sun to the 1971 Jan 0·0d position and assume that this is its position at Jan 0·0d for the year in question: it won't be more than about 1° out. Now set the node only to the Jan 0·0d position—that is, correct only for the years since 1971. Now we can see when eclipses are likely for the year simply by comparing the date scales which show when the Sun is at the Moon's node.

Take 1979 as an example. The Sun is at 279°, and the node at about 171°. The Sun will be near the Moon's node about March 7. Where will the Moon be? Setting the calculator completely for this date shows the Moon at about 91°, nearly 9 days after new moon. Try February 27. At 00h it is at about 342·5°, and past new moon by about 7 hours. So we can predict an eclipse of the Sun at 1979 February 26d 17h.

Now we can carry out the detailed solar and lunar position calculations. The total interval is 8 years 59 days 17 hours, giving the Sun's true longitude at 337·489°. The Moon's mean longitude is 335·981°, and when corrected comes to 337·593°, so at the time we have assumed the moment of central eclipse has passed. The Moon overtakes the Sun (or Earth's shadow) at the difference between their hourly rates, which is 0·508° per hour (which is also equal to 0·508′ per minute, or 30·5′ per hour). Dividing the difference in longitudes, 0·104°, by 0·508 gives 0·205 hours, or 12 minutes. The time of central eclipse will therefore be 17h 00m—12m = 16h 48m.

The mid-eclipse on this date is, in fact, predicted at 17h 22m, so we are pretty close.

The Moon's corrected latitude at this instant works out to 0° 55·8′, and the longitude 12 minutes earlier than our first estimate will be 337·481°. At the point on Earth from where the central eclipse will be seen, the eclipse will be taking place at local true noon, and hence the hour angle of the Moon and Sun will be zero. If we put H = 0 in our expression for correcting geocentric latitude, it will simplify. But remember that the simpler expression is only for the place on the track of the Moon's shadow where it is true local noon:

The latitude correction becomes HP Sin $(\phi - \delta)$.

Now, at the place of the central eclipse, the apparent latitude will be zero; the Moon will be truly centred on the Sun. So the amount by which we correct must equal the geocentric latitude. We can now say that HP Sin $(\phi - \delta) = 55{\cdot}8'$.

In this example, HP = 59·9′, and the declination of the ecliptic at longitude 337·5° is —9°. So Sin $(\phi + 9) = 55{\cdot}8/59{\cdot}9 = 0{\cdot}9316$. The angle which has the sine 0·9316 is 68·7°, so we have: $(\phi + 9) = 68{\cdot}7°$. Therefore,

$\phi = 59{\cdot}7°$, and this is the approximate latitude from which we can see the eclipse.

To find the approximate longitude on Earth all we need to do is work out the sidereal time at 1979 February 26d 16h 48m. This works out at 3h 11·4m. The RA of the Sun, from our diagram of the ecliptic, is 22h 34m, so the Greenwich hour angle is 4h 37·4m = 69° west. This will be the longitude of the place, putting it in northern Canada. The actual position is some 200 miles east of the position we have found, the errors being due to factors we have ignored, but we can see that the eclipse will be visible from the far north of the American continent.

Will it be total? To determine this we need to know the apparent sizes of the Sun and Moon. The semidiameter (radius) of the Sun is given by: $16{\cdot}025 + 0{\cdot}267 \text{ Cos } A_s$ arcmin.

The semidiameter of the Moon is: 0·272 HP. arcmin. In this example the semidiameter of the Sun works out to 16·18′, and the Moon's semidiameter is 16·3′. The Moon's disc is bigger than the Sun's, so the eclipse will be total.

Now let us take another example to show how a partial eclipse will appear from where we happen to be. This time we will start with data taken from the almanac to show how it can be used—but we could, of course, work it out for ourselves.

Partial solar eclipses

An eclipse of the Sun is predicted for 1975 May 11d 07h 38·5m. We can quickly work out the Sun's true longitude for this instant, and the Moon's will be the same. It is 50·021°. The polar co-ordinates are RA 3h 10m, Dec +17·7°. Using the calculator, or the mean motions from the tables, we find the perigee is at about 155·5°, and the ascending node at 241·7° so we can find the geocentric latitude of the Moon and correct it; it is 60·48′.

The other data we can now obtain includes the HP = 56·21′, the Sun's semidiameter, 15′ 51·7″; and the Moon's semidiameter, 15′ 17·4″.

From the HP and geocentric latitude we see that there is no terrestrial latitude at which the Moon's shadow can fall, because if β is larger than HP we need an angle with a sine greater than 1, and that is impossible. (Try it in our latitude correction formula.) Not only that, but the Moon's disc is smaller than the Sun's so it could only be an annular eclipse.

But what will be seen from Scotland, for example? Take a position on Earth of latitude 55° N and 5° W, and apply the full geocentric co-ordinate corrections for longitude and latitude.

The RA, as we have found, is 3h 10m, and the sidereal time works out to

22h 52·2m. The Greenwich hour angle is therefore 19h 42·2m = 295·55°. At longitude 5° W, the hour angle will be 5° smaller, 290·55°:

The correction for longitude at 7h 38·5m UT comes to +31·7′, and the latitude correction to −40·42′, giving an apparent position of longitude 50° 33′, latitude +20·06′.

Now we can draw a plan of the ecliptic to a large scale, for example 1mm per arcmin, showing longitudes between about 49° and 51°. Draw

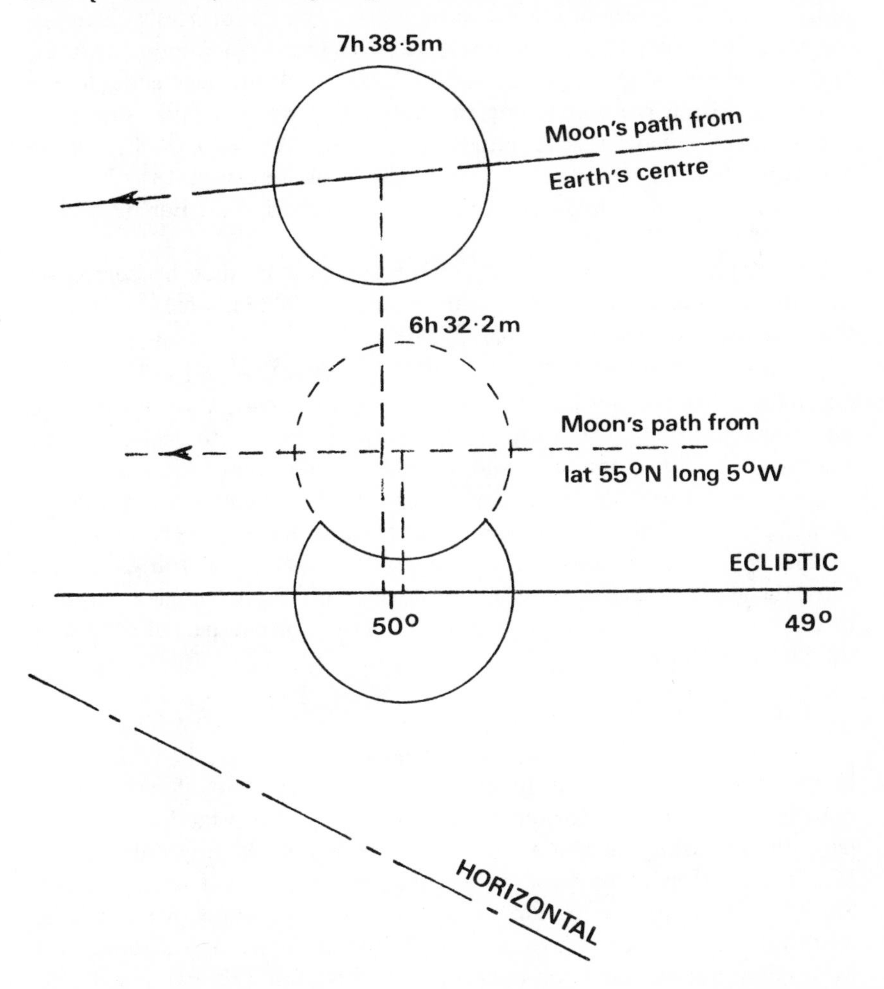

Fig 73 Diagram of the eclipse of the Sun 1975 May 11d, showing how the Moon passes above the Sun as seen from the Earth's centre, but causes a partial eclipse in Scotland, as shown below.

the Sun to scale (radius 15′ 52″) at longitude 50° 1·3′. Now draw a vertical line at the same longitude, measure 60·5′ north, and draw the Moon to scale (radius 15′ 17″) to show the geocentric appearance: the Moon misses comfortably, as is shown in Fig 73.

It is useful, particularly for plotting lunar eclipses as we shall in a moment, to be able to plot the slope of the ecliptic. We can do this over a short distance like this by assuming a straight line between two position plots. From the centre of the Earth only (*not* from its surface) we can say the Moon's latitude changes at a rate of 3 Cos $(\lambda_m - \Omega)$ arcmin per hour, and near the node at eclipse times this comes to 3′/hr, near enough. We know the Moon is approaching the descending node, so that one hour earlier it was at geocentric longitude 50° 1·3′—0·549° = 49° 28·4′, and the geocentric latitude was 60·48′+3 = 63·48′. Now we can plot the Moon's centre at this position and draw in its path as seen from the Earth's centre, as shown in Fig 73.

Unfortunately, we must work out the apparent latitude by correcting this all over again for the new hour angle at 6h 38·6m, which is 15° less than for the first position plotted, 275°·55.

This gives an apparent position of longitude 50° 2′, latitude 20·5′ at 6h 38·6m. Now we can plot the Moon's apparent path as seen from the observer's position on the Earth. We draw the Moon to scale, centred exactly at the Sun's longitude, and find the apparent central eclipse occurs 3·2′, equivalent to 6·3 minutes earlier than our 6h 38·6m position, that is, at 6h 32·2m. To finish off the prediction of what the eclipse will look like, we can estimate the angle of the ecliptic to the horizon from our globe and show the approximate orientation on the diagram. By turning the diagram to the horizontal, we see how Scotland's early morning partial eclipse of the Sun will appear.

Lunar eclipses

Eclipses of the Moon are of little interest to the professional astronomer, but they are fascinating for the amateur to watch. When the Earth is only partially obscuring the Sun as seen from the Moon, the brightness of the Moon is scarcely diminshed, so we will concentrate on total eclipses. Since the Moon is entering the cone of the Earth's shadow, which can be well over twice the semidiameter of the Moon at the Moon's distance, the Moon need not be exactly on the ecliptic as shown (not to scale) in Fig 71. The ecliptic longitude of the centre of the Earth's total shadow, or umbra, will be the same as the Earth's, so we can plot its position on the ecliptic in much the same way as we did for the Sun. But since the eclipse will be visible from any point on Earth where the Moon is above the horizon, and because the parallax effect on the Moon's position due to our position on the Earth's surface will apply equally to the position of the shadow, we need only work out the geocentric conditions. We can therefore plot the Moon's path through the umbra, using the change in latitude of 3′/hr near the node, to determine the exact moment of central eclipse.

We will ignore the small effect on the diameter of the Earth's shadow due to the change in the Earth–Sun distance during the year, so the expression for the semidiameter of the Earth's shadow at the distance of the Moon is:

HP—16 arcmin.

We use the calculator for our initial estimate, just as before, or we can use a diagram of the nodes' motion. But this time the Moon is exactly opposite the Sun, that is, it is full moon.

As an example, try the year 2000. From the calculator, it can be seen that the Sun reaches the descending node about January 23, and the ascending node at about July 15. When is full moon in January? At January 23 we find the Moon is about 2 days 4 hours passed full. Try

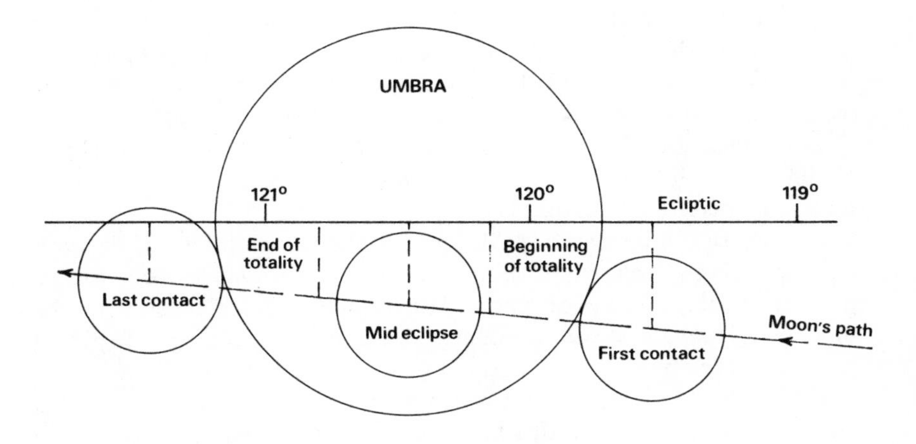

Fig 74 This eclipse of the Moon is visible from America and Britain in 2000 January 21.

again for January 20d 20h. Now we have 6 hours to go, so our initial estimate is 2000 January 21d 2h.

The Earth's true longitude, and hence the centre of the umbra, works out to 120·331°. The Moon's true longitude comes to 118·770°, so it still has 1·561° to go, overtaking the Earth's shadow at 0·508° per hour. The mid-eclipse will therefore be at 5h 4m UT. The longitude of the Moon and umbra will then be 120·457°. The Moon's latitude works out at −18·5′, and from the calculator we see it is approaching the ascending node. We can now work out the semidiameters and plot a diagram. The umbra is 43·6′ radius, and the Moon 16·2′. Fig 74 shows the Moon at the middle of the eclipse.

We can plot the Moon's path towards the node by plotting any point 0·549° (33′) in longitude from the centre of the umbra (one hour's motion) and changing the latitude by 3′. For example, we can plot a point at longitude 121° 00·5′, latitude −15·5′, and draw the Moon's path. Now, using only the *relative* rate of the Moon's motion to avoid having to replot the position of the umbra, we can measure the difference in longitude when the Moon just touches the shadow as it enters and leaves the umbra, called the first and last contacts, and the relative position when the Moon is totally immersed in the shadow on its way in and out, the beginning and end of totality. We do this by drawing the Moon's disc to the same scale, as shown in Fig 74 for the first and last contact positions only. By dividing the difference in longitude by the relative rate of the Moon's motion (0·508°/hr) we find from the drawing the following times:
First contact, 3h 15·7m; beginning of totality, 4h 28·6m; mid-eclipse, 5h 4m; end of totality, 5h 43·4m; last contact 6h 58·2m. The duration of totality is therefore 75 minutes.

The eclipse will be seen from Britain and the United States. If we work out the Moon's declination we find it is 20° north, and hence it will be overhead at this latitude on Earth. The hour angle is 73°W, so it will be on the meridian and it will be midnight local time at this longitude, which is close to New York.

This is the Moon's motion considerably simplified, but quite detailed enough. You can work with simple basic data, or refine it to your own requirements, and with practice, you can work up quite a speed.

EXERCISES ON CHAPTER 9

1 Calculate the position of the Moon as seen from latitude 60°N, longitude 150°W (in Alaska) at 1978 December 13d 2h 35m Alaskan Standard Time (UT−10h). Draw a diagram and plot Aldebaran's position on the same diagram.
2 Can the Sun ever rise or set totally eclipsed?
3 Draw to scale on the ecliptic the discs of the Sun and Moon as seen from

Cornwall, at latitude 50°15′N, longitude 5°W, at 1999 August 11d 11h.

4 The Moon was particularly worth watching at 1972 March 19d 19h. Why?

5 On what date is the first eclipse of the Sun in 1984? Will totality last a particularly long time?

6 One evening from a place at latitude 50°N or so, you see a thin crescent Moon. Is it just after or just before new Moon? If the crescent appears to be 'on its back', what is the approximate date and what is the approximate ecliptic latitude of the Moon?

7 What was the age of the Moon in 1789, on April 28d?

8 Using only the instruments you have made find the Pacific Standard Time of moonset from Los Angeles (say latitude 34°N, longitude 120°W) at 1975 November 24d.

9 By setting the ecliptic latitude scale (disc 6) of the calculator so that the 'Moon' mark points at the ecliptic longitude of a star, and setting the ☊ mark on disc 5 opposite the ecliptic latitude of that star on the inner disc, you can read on the ecliptic longitude scale opposite ☊ the longitude that the ascending node must have (two positions for the two position of latitude) for the Moon to occult that star, approximately. When will the Moon next occult Antares after 1975? Will it be well placed?

Postscript

The aim of this book has been to show just how much practical astronomy can be carried out at virtually no expense by any amateur with only a modicum of practical ability and an elementary amount of mathematics. There is no way the slog of some of the experiments can be reduced and the reasonable accuracy of the result maintained, but even the more formidable procedures we have examined take much longer to explain than to do when you have had some practice.

The important word is practice. The beauty of simple positional astronomical experiments is that you can carry them out to the degree of accuracy that suits your purposes. Even walking home on a starlit night if you want a pleasant and informative way to while away the journey, you can carry out the navigational exercises in your head (simplified considerably, of course!) you can work out the time, and, most important, you should be unable to lose your sense of direction.

The amateur who has developed a taste for experiments of this type can continue to broaden his horizons with a camera, or a simple telescope, and hints have been dropped from time to time of what you can do with simple instruments other than just gazing at the sky.

The home-made calculators and instruments which have been described in these pages will become invaluable rapid references and accessorics for the telescopic user, and with their use will come an awareness of what the sky is doing all the time above your head. They will save you much wasted time looking for faint objects which you should now be able to pinpoint both amid the surrounding stars and in relation to your own position on the Earth.

Finally, much of what we have done retraces the steps of the ancient astronomers, and this sort of study certainly helps you to appreciate their achievements.

So, rather than make this the end, why not make it a start to a new way of looking at this, the oldest of the sciences and the one about which Man still has the most to learn.

Appendix A
Answers to Exercises

Some of your answers may differ slightly from those given and still be perfectly satisfactory. The main thing is to get the method right!

Chapter 1
1 Shadow points north; shortest at noon; east; 45°; 6h; at an equinox.
2 Always 90°; lat 60°N; at the equator.
3 Lat 90°N; Lat 0°; Alt 45°.

Chapter 2
1 9·25°W.
2 Sunset is at 18h 0m.
3 Moon is about 12° slow on Sun for each day after new; other dials not easy to use.
4 10h 39½m.

Chapter 3
1 00h 04m.
2 RA 0h, Dec 0° approx.
3 About December 7d.
4 15h 15m.
5 June–July.
6 7h 30m.

Chapter 4
1 6°.
2 δ Orionis is almost on the equator in Orion's Belt so rises and sets east and west. Any star is south when its RA = ST.
3 Cygnus; Auriga.

Chapter 5
1 3h 55m.
2 17h 35m.
3 Athens (Lat 38°N, Long 23°45′E).
4 73°N.
5 February 25d.
6 Use the altitude scale as a protractor and measure on the globe, 32°.
8 Azimuth 131°; December 21.
9 4¾ hours, easily seen.
10 Latitude 74½°N.

Chapter 6
1 Three.
3 (a) 10h 54m, +10°; (b) 14h 05m, +10°.

Chapter 7
1 (Pole-centred diagram). Procyon, RA 7h 39m, Dec +4·4° by this calculation—see true value.
2 (Pole-centred diagram), 15h 07·4m.
3 (Zenith-centred) Alt 67°, Az 39°.
4 (Zenith-centred diagrams). Lat 42° 48′N, Long 13° 00′W.

Chapter 8
2 Jupiter and Saturn.
3 1975 Aug. 27d; at inf. conj., planet is *closer* to Earth–no transit in 1975.
4 RA 1h 03m; Dec+5°.
5 Eastern; evening star; no.
6 154 AU, 1,910 years: 308 AU, 5,405 years; 615 AU, 15,250 years. 1975 July 11d 12h. 34·5m.
7 Longitude 94° approximately.
8 65·6 km.

Chapter 9
1 Occultation of Aldebaran.
2 Often.
3 Total eclipse of Sun.
4 Moon occulted several of the Pleiades.
5 May 31. Annular.
6 Passed new, close to spring equinox. Moon at Lat +5°.
7 3 days.
8 11h 16m.
9 1987 July 8d 12h (approx) visible in eastern hemisphere.

Appendix B
Mathematical Methods and Tables

Some of the calculations explained in detail in the text are summarised in this appendix, and tables to calculate sidereal time and planetary positions are given, together with a list of symbols and abbreviations used in astronomical work.

Symbols and Abbreviations

° ′ ″	Degrees, minutes, and seconds of arc (arcmin and arcsec).
α	Right ascension (RA).
δ	Declination (Dec).
β	Geocentric latitude.
λ	Geocentric longitude.
L	Heliocentric longitude
ϕ	Terrestrial latitude.
♈	First point of Aries (ie RA 00h 00m or 0° longitude on the equator).
☊	Ascending node (and its ecliptic longitude).
☋	Descending node (and its ecliptic longitude).
ϖ	Longitude of perihelion.
Γ	Longitude of perigee.
a	Semi-major axis of an orbit (usually in AU or km).
AU	Astronomical unit.
e	Eccentricity.
GMT	Greenwich Mean Time.
UT	Universal time (= GMT for our purposes).
GST	Greenwich sidereal time.
LST	Local sidereal time.
h m s	hours, minutes, and seconds of time.
H	Hour angle (HA) measured westwards from the meridian.
HP	Horizontal parallax.
i	Inclination of orbit to ecliptic.
n	Mean daily motion of planet around its orbit.
h	altitude.
Az	azimuth.
D	Elongation of Moon.
E	Elongation of planet.
A	Moon's anomaly.

Solutions to astronomical triangles (see Fig 46)

Given hour angle, declination and latitude, find altitude and azimuth:

$$\text{Sin h} = \text{Cos H Cos } \phi \text{ Cos } \delta + \text{Sin } \phi \text{ Sin } \delta$$

$$\text{Cos Az} = \frac{\text{Sin } \delta - \text{Sin } \phi \text{ Sin h}}{\text{Cos } \phi \text{ Cos h}}$$

Given altitude, azimuth, and latitude, find hour angle and declination:

$$\text{Cos H} = \frac{\text{Sin h} - \text{Sin } \phi \text{ Sin } \delta}{\text{Cos } \phi \text{ Cos } \delta}$$

$$\text{Sin } \delta = \text{Cos } \phi \text{ Cos Az Cos h} + \text{Sin } \phi \text{ Sin h}$$

There are two answers for each angle in the cases of hour angle and azimuth, but these values will always be in the order of 180° different, since HA is measured from the meridian, and azimuth from the north point. Negative altitudes show angle beneath observer's horizon (twilight calculations).

Arithmetical solution to plane triangles

RIGHT-ANGLED TRIANGLES (FIG 75, LEFT)

The side of the triangle opposite the right angle is the hypotenuse, h. Then:

$$\text{Sine A} = \frac{a}{h} \quad \text{Cosine A} = \frac{b}{h} \quad \text{Tangent A} = \frac{a}{b}, \text{ hence } \frac{\text{Sin A}}{\text{Cos A}} = \text{Tan A}.$$

$$\text{Sine B} = \frac{b}{h} \quad \text{Cosine B} = \frac{a}{h} \quad \text{Tangent B} = \frac{b}{a}, \text{ hence } \frac{\text{Sin B}}{\text{Cos B}} = \text{Tan B}.$$

Note that angles A and B summate to 90°, and that $a^2 + b^2 = h^2$.

ANY PLANE TRIANGLE (FIG 75)

To find the third side, given two sides and the angle between them,

$$a^2 = b^2 + c^2 - 2bc \text{ Cos A}$$

Given three sides; to find an angle:

$$\text{Cos A} = \frac{b^2 + c^2 - a^2}{2bc}$$

In other cases:

$$\frac{a}{\text{Sin A}} = \frac{b}{\text{Sin B}} = \frac{c}{\text{Sin C}}$$

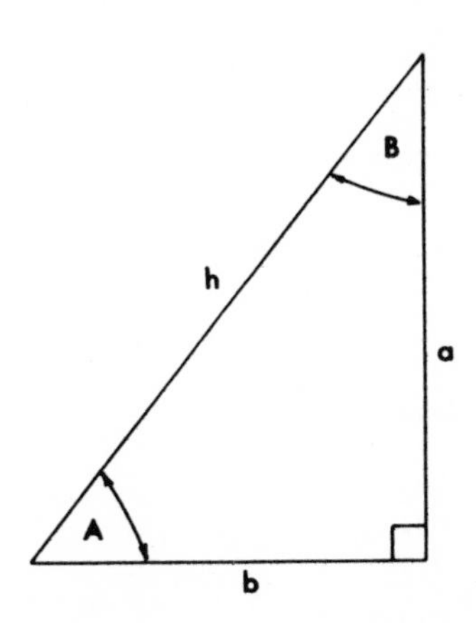

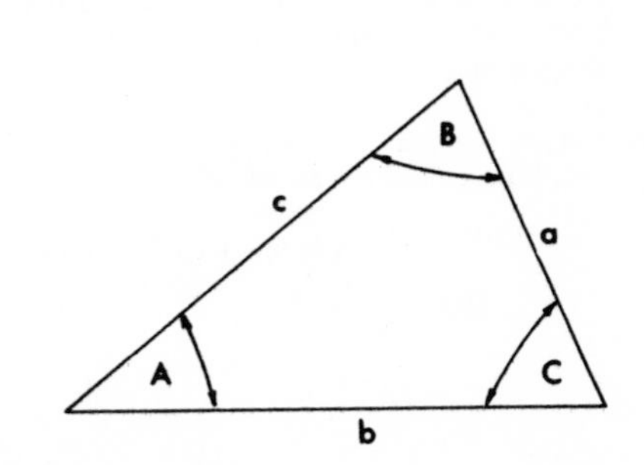

Fig 75

Sidereal time at 00h00m GMT throughout the year

Date		h m	Date		h m	Date		h m	Date		h m
Jan	1	6 40.9	Apr	3	12 43.6	July	4	18 46.4	Oct	4	0 49.1
	5	6 56.7		7	12 59.4		8	19 02.1		8	1 04.9
	9	7 12.5		11	13 15.2		12	19 17.9		12	1 20.6
	13	7 28.2		15	13 31.0		16	19 33.7		16	1 36.4
	17	7 44.0		19	13 46.7		20	19 49.4		20	1 52.2
	21	7 59.8		23	14 02.5		24	20 05.2		24	2 07.9
	25	8 15.5		27	14 18.3		28	20 21.0		28	2 23.7
	29	8 31.3									
Feb	2	8 47.1	May	1	14 34.0	Aug	1	20 36.8	Nov	1	2 39.5
	6	9 02.9		5	14 49.8		5	20 52.5		5	2 55.2
	10	9 18.6		9	15 05.6		9	21 08.3		9	3 11.0
	14	9 34.4		13	15 21.3		13	21 24.1		13	3 26.8
	18	9 50.2		17	15 37.1		17	21 39.8		17	3 42.6
	22	10 05.9		21	15 52.9		21	21 55.6		21	3 58.3
	26	10 21.7		25	16 08.7		25	22 11.4		25	4 14.1
				29	16 24.4		29	22 27.1		29	4 29.9
Mar	2	10 37.5	June	2	16 40.2						
	6	10 53.3		6	16 56.0	Sept	2	22 42.9	Dec	3	4 45.6
	10	11 09.0		10	17 11.7		6	22 58.7		7	5 01.4
	14	11 24.8		14	17 27.5		10	23 14.5		11	5 17.2
	18	11 40.6		18	17 43.3		14	23 30.2		15	5 32.9
	22	11 56.3		22	17 59.1		18	23 46.0		19	5 48.7
	26	12 12.1		26	18 14.8		22	0 01.8		23	6 04.5
	30	12 27.9		30	18 30.6		26	0 17.5		27	6 20.3
							30	0 33.3		31	6 36.0

Times given in the table are for any year two years after a leap year. For other hours, days and years, correct these times as follows:

At t hours	add
3	0·5m
6	1·0m
9	1·5m
12	2·0m
15	2·5m
18	3·0m
21	3·5m
24	3·9m

For x days,	add
1	3·9m
2	7·9m
3	11·8m

Years	add
1 yr after leap year	+1m
2 yrs after leap year	0m
3 yrs after leap year	−1m
Leap year, Jan 1–Feb 28	−2m
Remainder of leap year	+2m

Sidereal time Feb 29d 00h: 10h 31·4m

Example: What is GST at 1972, June 12d 16h 36·5m?
GST at 00h = 17h 11·7m on June 10d
add, for two days 7·9m
add, for 16·5h 2·75m
add, for leap year 2·00m
add, at time given 16h 36·5m

34h 00·9m. Reject multiples of 24h; GST = 10h 00·9m.

Sun's declination, and the equation of time

Date	Declination °	Equn. of time (minutes)	Date	Declination ° ′	Equn. of time (minutes)
Jan 1	−23 01	− 3.3	Jul 4	+22 54	− 4.2
5	−22 38	− 5.3	8	+22 30	− 4.9
9	−22 07	− 7.0	12	+22 00	− 5.5
13	−21 30	− 8.7	16	+21 24	− 5.9
17	−20 46	−10.1	20	+20 42	− 6.3
21	−19 56	−11.3	24	+19 54	− 6.4
25	−19 00	−12.4	28	+19 01	− 6.4
29	−17 58	−13.1	Aug 1	+18 04	− 6.2
Feb 2	−16 51	−13.7	5	+17 01	− 6.0
6	−15 39	−14.1	9	+15 54	− 5.5
10	−14 23	−14.3	13	+14 43	− 4.7
14	−13 04	−14.2	17	+13 28	− 4.1
18	−11 41	−14.0	21	+12 10	− 3.1
22	−10 15	−13.6	25	+10 49	− 2.1
26	− 8 46	−13.0	29	+09 25	− 1.0
Mar 2	−07 15	−12.2	Sep 2	+07 58	+ 0.2
6	−05 43	−11.3	6	+06 30	+ 1.6
10	−04 09	−10.4	10	+05 00	+ 3.0
14	−02 35	− 9.3	14	+03 28	+ 4.3
18	−01 00	− 8.1	18	+01 56	+ 5.8
22	+00 35	− 7.0	22	+00 22	+ 7.2
26	+02 09	− 5.8	26	−01 11	+ 8.6
30	+03 43	− 4.5	30	−02 45	+ 9.9
Apr 3	+05 15	− 3.4	Oct 4	−04 18	+11.2
7	+06 47	− 2.2	8	−05 50	+12.4
11	+08 16	− 1.1	12	−07 21	+13.4
15	+09 43	− 0.1	16	−08 50	+14.4
19	+11 08	+ 0.8	20	−10 17	+15.2
23	+12 29	+ 1.7	24	−11 42	+15.7
27	+13 47	+ 2.4	28	−13 04	+16.2
May 1	+15 02	+ 2.9	Nov 1	−14 23	+16.4
5	+16 13	+ 3.4	5	−15 38	+16.3
9	+17 19	+ 3.7	9	−16 49	+16.2
13	+18 21	+ 3.9	13	−17 56	+15.8
17	+19 18	+ 3.7	17	−18 57	+15.1
21	+20 09	+ 3.6	21	−19 53	+14.2
25	+20 56	+ 3.3	25	−20 43	+13.1
29	+21 36	+ 2.8	29	−21 28	+11.7
Jun 2	+22 10	+ 2.2	Dec 3	−22 05	+10.2
6	+22 38	+ 1.5	7	−22 36	+ 8.6
10	+23 00	+ 0.7	11	−22 59	+ 6.9
14	+23 16	− 0.1	15	−23 16	+ 4.9
18	+23 24	− 0.9	19	−23 25	+ 3.0
22	+23 27	− 1.7	23	−23 26	+ 1.1
26	+23 22	− 2.7	27	−23 20	− 0.9
30	+23 11	− 3.5	31	−23 07	− 2.9

Figures applicable to a year two years after a leap year (or two years before) and are for the Sun at noon, ie 12h 00m GMT.
Equation of time: positive, sundials fast; negative, sundials slow.
Apparent (sundial time) = Mean time + equation of time.

MEAN ELEMENTS OF PLANETARY ORBITS, 1971 JANUARY 0.5d. *All values of longitude are heliocentric*

Planet	Mean longitude of date (L) (degrees)	Add, annual variation in L (degrees)	Longitude of perihelion (ϖ) (degrees)	Sidereal mean daily motion, n (degrees)	Mean distance a (AU)	Eccentricity e	115e (for eq of c)	Longitude of ascending node, ☊ (degrees)	Inclination of orbit to ecliptic i (degrees)
Mercury	101·700	53·717	77·004	4·09234	0·38710	0·20563	23·6	47·988	7·004
Venus	130·206	224·792	131·163	1·60213	0·72333	0·00679	0·78	76·418	3·395
Earth	99·504	* −0·239	102·441	0·98561	1	0·01672	1·92	—	—
Mars	203·960	191·286	335·525	0·52403	1·52359	0·09338	10·7	49·333	1·850
Jupiter	233·762	30·341	13·854	0·08309	5·20280	0·04845	5·56	100·155	1·305
Saturn	55·233	12·227	92·480	0·03346	9·53884	0·05564	6·4	113·404	2.489
Uranus	188·586	4·296	170·190	0·01173	19·1819	0·04724	5·43	73·853	0·772
Neptune	241·120	2·197	44·371	0·00598	30·0579	0·00858	0·985	131·461	1·771
Pluto	197·528	1·467	223·831	0·004	39·457	0·24852	28·5	109·934	17·144

* N.B. Retrograde

Summary of data and formulae for planetary and lunar position calculation

PLANETS (AND SUN)

Mean longitude calculated from data in table of mean elements of planetary orbits 1971 January 0·5 days. Correct for equation of centre:

$$\frac{180}{\pi} \times 2e \text{ Sin}(L - \varpi)$$

Take $\frac{180}{\pi} \times 2$ to equal 115, then values of 115e are given in table.

Plot planet's orbit, and that of Earth to scale, mark positions and project from Earth to find geocentric longitudes. Geocentric latitude is:

$$\beta = \frac{ai}{S} \text{ Sin}(L - \Omega)$$

where S is distance Earth to planet in AU, measured on diagram. Sun's geocentric longitude = Earth's heliocentric longitude — 180°.

MOON

Mean sidereal period: 27·32166 days
Mean synodic period: 29d 12h 44m 3s
Equatorial diameter: 3476 km (2160 miles)
Mean distance from Earth: 384,400 km (238,868 miles, or 0·002569 AU)
Mean inclination of orbit to ecliptic: 5·1453° (5° 8′ 43″)
Mean eccentricity: e = 0·05490 (115e = 6·3)
Calculate the mean longitudes of Earth, Moon, the Moon's perigee and node from the following mean elements for 1971 January 0d 00h:

	Mean Geocentric longitude, 1971 Jan. 0·0d	*Sidereal mean hourly motion*	*Sidereal mean daily motion*	*Net mean annual motion*
Earth:	99·011°	0·0411°	0·9856°	—0·239° (retro)
Moon:	314·1601°	0·549°	13·1764°	129·385°
Perigee:	343·199°	—	0·1114°	40·662°
Node:	326·009°	—	—0·05295° (retro)	—19·328° (retro)

NET CHANGES IN MOON'S MEAN LONGITUDE AT 50 DAY INTERVALS:

To simplify the calculation of mean longitude at the 13·1764° per day rate, the *changes* in mean longitude listed below can be added to the Jan. 0·0 value and only the remaining days' motion need be calculated:

For 50 days after Jan 0·0d add 298·820° to mean value for the year at Jan 0·0d
,, 100 ,, 237·640° ,,
,, 200 ,, 115·279° ,,
,, 300 ,, 352·919° ,,

SUMMARY OF FORMULAE AND DATA FOR LUNAR POSITION, AND ECLIPSE CALCULATIONS:

Apply the following longitude corrections to λ_m, A and D:
Evection: 1·3 Sin (2D $-$A)
Annual inequality: $-$0·2 Sin A_s, where A_s is Sun's anomaly
Equation of centre: 6·3 Sin A + 0·2 Sin 2A
Variation: 0·7 Sin 2D
Moon's geocentric latitude, β (+ is north of ecliptic): 5·15 Sin($\lambda_m - \Omega$) degrees
Latitude inequality (correction to β): 10 Sin(2D $-\lambda_M + \Omega$) arc min.
Correction of geocentric lunar position to account for observer's position:
To correct the geocentric longitude, *subtract* from λ_M the following amount:

$$HP \frac{\cos \phi \sin H}{\cos \delta} \text{ arcmin.}$$

(Note that this and the following term are in arcmin.)
To correct the geocentric latitude, *subtract* from β_M

$$HP (\sin \phi \cos \delta - \cos \phi \sin \delta \cos H) \text{ arcmin}$$

Note: D and A are measured westwards from the Moon. H is measured westwards from the observer's meridian.

USEFUL DATA

Moon's mean horizontal parallax: HP = 57·04 +3·11 Cos A arcmin
Moon's semidiameter: 0·272 HP arcmin
Sun's semidimeter: 16·025 + 0·267 Cos A_s arcmin
Semidiameter of Earth's shadow at Moon's distance: HP $-$16 arcmin
Geocentric longitude of Sun's perigree: 282·442° (1971, but ignore changes)
Moon "overtakes" the Sun or the Earth's shadow at: 0·508° per hour (30·5 arcmin/hr)
Change in Moon's *geocentric* latitude: 3 Cos ($\lambda_M - \Omega$) arcmin/hr. (i.e. at eclipse times, when Moon is near node, the geocentric latitude changes at 3′/hr.)

Moon's distance from Earth centre: $\frac{21{,}926{,}700 \text{ km}}{HP'}$ or $\frac{13{,}624{,}619 \text{ miles}}{HP'}$

Number of days through the year

(to assist in lunar and planetary position work)

When the date is January 1 add 1 day's mean daily motion to λ at Jan. 0d.

1 February	32	
1 March	60	(add one more day in leap years after this date).
1 April	91	
1 May	121	
1 June	152	
1 July	182	
1 August	213	
1 September	244	
1 October	274	
1 November	305	
1 December	335	

Note: Subtract 0·5d when starting from 1971 Jan 0·5d.

LEAP YEAR

Leap years occur in general every fourth year. When the year is divisible by 4 exactly there is a leap year, except when the year is also the end of a century. For example, 1900 although exactly divisible by 4, was not a leap year. The need to omit a leap year at the end of the century arises because there are 365·24219 days in a tropical (equinox to equinox) year. Even this omission every century is not accurate enough, so that when the century ends with a year divisible by 400 it is made a leap year after all! The year 2000 will be a leap year.

Starting from 1971 Jan 0d when our tables list the positions of the Sun, Moon, Earth, and the other planets, we must add one day's mean daily motion for each leap year (after February 29).

In the year:	1972	'76	'80	'84	'88	'92	'96	2000	'04	'08	'12	'16
Add (days);	1	2	3	4	5	6	7	8	9	10	11	12

Example How many years' and days' mean daily motion should be used on 1996 January 18? You must add 25 years' motion, and 6 *days* (we have not yet reached February 29 in this year), then the 18 days in January.

Some star names, magnitudes and positions (1971)

Star	Name	Magnitude	Right ascension	Declination
α And	Alpheratz	2·1	0h 06·9m	+28° 56′
β Cas	Caph	2·4	0h 07·6m	+58° 59′
α Cas	Schedar	2·3	0h 38·8m	+56° 23′
α Eri	Achernar	0·6	1h 36·6m	−57° 23′
α UMi	Polaris	2·1	2h 04·1m	+89° 04′
α Ari	Hamal	2·2	2h 05·5m	+23° 20′
o Cet	Mira	2–10	2h 17·9m	−3° 07′
β Per	Algol	2·0–3·0	3h 06·3m	+40° 51′
α Per	Mirfak	1·9	3h 22·2m	+49° 46′
η Tau	Alcyone	3·0	3h 45·8m	+24° 01′
α Tau	Aldebaran	1·1	4h 34·3m	+16° 27′
β Ori	Rigel	0·3	5h 13·1m	−8° 14′
α Aur	Capella	0·2	5h 14·5m	+45° 58′
γ Ori	Bellatrix	1·7	5h 23·6m	+16° 19′
α Ori	Betelgeuse	0·0–1·0	5h 53·6m	+7° 24′
α Car	Canopus	−0·9	6h 23·3m	−52° 40′
α CMa	Sirius	−1·6	6h 43·9m	−16° 40′
α Gem	Castor	1·6	7h 32·7m	+31° 57′
α CMi	Procyon	0·5	7h 37·8m	+5° 18′
β Gem	Pollux	1·2	7h 43·5m	+28° 06′
α Hya	Alphard	2·2	9h 26·2m	−8° 22′
α Leo	Regulus	1·3	10h 06·8m	+12° 07′
β UMa	Merak	2·4	11h 00·1m	+56° 32′
α UMa	Dubhe	1·9	11h 01·9m	+61° 54′
β Leo	Denebola	2·2	11h 47·6m	+14° 44′
α CVn	Cor Caroli	2·9	12h 54·7m	+38° 28′
ζ UMa	Mizar	2·4	13h 22·8m	+55° 05′
α Vir	Spica	1·2	13h 23·7m	−11° 01′
α Boo	Arcturus	0·2	14h 14·3m	+19° 20′
α CrB	Gemma	2·3	15h 33·5m	+26° 49′
α Sco	Antares	1·2	16h 27·6m	−26° 22′
α Lyr	Vega	0·1	18h 36·0m	+38° 45′
β Cyg	Albireo	3·2	19h 29·5m	+27° 54′
α Aql	Altair	0·9	19h 49·4m	+8° 47′
α Cyg	Deneb	1·3	20h 40·4m	+45° 11′

Further reading

The following list of publications and books will provide further information on some of the subjects covered in this book, and yearly data on astronomical phenomena. Most of these publications are available in most parts of the English speaking world.

Cousins, Frank W. *Sundials* (London: John Baker 1969).

Tricker, R.A.R. *The Paths of the Planets* (London, Mills and Boon 1967).

Sidgwick, J. B. *Amateur Astronomer's Handbook* and *Observational Astronomy for Amateurs*. (London, Faber 1971.)

Woolard, E. A. and Clemence G. M. *Spherical Astronomy* (New York, Academic Press).

Handbook of the British Astronomical Association published by the B.A.A., Burlington House, Piccadilly, London W1V 0NL.

The Astronomical Ephemeris (Published in London annually by HMSO and in the United States as the *American Ephemeris and Nautical Almanac* by the US Government Printing Office, Washington).

Norton, A. P. and Gall Inglis, J. *Norton's Star Atlas and Telescopic Handbook* (Edinburgh, Gall and Inglis).

The Observer's Handbook (Royal Astronomical Society of Canada), 252 College Street, Toronto 2B, Ontario.

Whipple, F. L. and Lundquist, C. *Smithsonian Astrophysical Observatory Star Atlas* (Cambridge Mass., MIT).

Sky and Telescope (published monthly by Sky Publishing Corp., Cambridge Mass). Many other periodical publications are available in the United States covering various items as included in the Astronomical Ephemeris from the local astronomical societies.

Index

Abbreviations of astronomical terms, 192
Albireo, 73
Alcor, 61
Aldebaran, 64
Algol, 65
Altair, 73
Altitude, 10
  calculation of, by stereographic projection 131;
  measurement of with quadrant, 101;
  of Polaris, 21;
  scale for celestrial globe, 89
Analemmatic sundial, 36
Andromeda, constellation of, 62
Annual inequality, Moon's, 167
Anomaly, Moon's, 166;
  Planets' 155
Antares, 72
Aphelion, 24
Aquarius, constellation of, 76
Aquilla, constellation of, 74
Arctic and antarctic circles, 14
Arcturus, 70
Aries, constellation of, 62;
  First Point of, 20, 63;
  transit of First Point of, 49
Astronomical triangle, 120;
  mathematical solutions of, 192
Astronomical Unit, 139
Auriga, constellation of, 65
Astrolabe, making and using, 102-116
Axis, Earth's, 13;
  Moon's rotation on, 163
Azimuth, 11
  calculation of by stereographic projection, 124-127;
  quadrant scale of, 101;
  scale for celestial globe, 87

Betelgeuse, 63
Binoculars, finding faint objects with, 62;
  projection of Sun's image with, 159
Bode's Law, 149
Bootes, constellation of, 70

Calculators: Moon's position, 170;
  planetary position, 146;
  sidereal time, 55, 86-88, 112
Cancer, constellation of, 68;
  tropic of, 14
Canes Venatici, constellation of, 70
Canis Major, constellation of, 66
Canis Minor, constellation of, 66
Capella, 65
Capricorn, constellation of, 76;
  tropic of 14
Capuchin Sundial, 41
Cassiopeia, constellation of, 60;
  sidereal time by 52
Castor, 68
Celestial globe, making and using, 82-92
Celestial sphere, 17

Cepheus, constellation of, 62
Chords, measuring angles by, 98
Co-declination, 119
Co-latitude, 120
Coma Berinices, constellation of, 70
Conjunction, 140
Constellations, 57-77
Corona, Sun's 179
Corona Borealis, constellation of, 71
Corvus, constellation of, 70
Cosine formula for solution of plane triangles, 10, 193
Cross staff, 94-98
Culmination, 19
Cygnus, constellation of, 73

Declination, 46;
  Sun's, effect on sundials, 38;
  Sun's, table of, 195
Delphinus, constellation of, 76
Deneb, 73
Draco, constellation of, 62
Dubhe, 60

Earth, axial inclination of, 14;
  daily rotation of, 9, 13, 48;
  distance from Sun, 24;
  equator of, 13;
  orbit of, 158,
  orbit, measuring shape of, 158;
  poles, 13;
  tropics, 14
Eclipses, 179-188
Ecliptic, Sun's movement around and inclination of, 26;
  plane of Earth's orbit, 140
Ecliptic latitude and longitude, 142
Elongation of planet from Sun, 141
Equator, celestial, 21;
  galactic, 66, 76;
  terrestrial, 13
Equation of the centre, Moon, 166, 168, 198;
  planets, 155
Equation of time, 26;
  table of, 195
Equinox, spring and autumn, 16;
  precession of, 19.
Evection, Moon's, 167, 198

Fomalhaut, 62

Galaxy, 66
Gemini, constellation of, 67
Globe, making a celestial, 82;
  use of celestial, 90
Greek, alphabet, 58
Greenwich Mean Time, 10;
  meridian, 17;
  Royal Observatory, 17

Hamal, 62
Hand, as a sundial, 35
Heliocentric co-ordinates, conversion to R.A. and Dec., 144
Hercules, constellation of, 72
Hour Angle, 50
Hour Circle, 47
Horizon, declination of observer's, 105
Horizontal parallax, 176
Hyades, 64
Hydra, constellation of, 68

Inferior conjunction, 141
Inferior planets, 140

Kepler's laws of planetary motion, 139

Latitude, ecliptic, geocentric and heliocentric, 142;
  heliocentric, calculation of, 156;
  Moon's ecliptic, 168;
  Moon's ecliptic, corrections for observer's position, 177;
  terrestrial, 17
Lepus, constellation of, 63

Leo, constellation of, 68;
the "Sickle", rising at different latitudes, 23
Light year, 16
Longitude, ecliptic, geocentric and heliocentric, 142;
Moon's ecliptic, 167;
Moon's ecliptic, corrections for observer's position, 178;
planets' heliocentric, 142, 152;
terrestrial, 17
Lyra, constellation of, 73;
ε Lyrae, the double double star, 73

Magnitudes, stellar, 58;
table of, 200
Mars, orbit of, 152
Merak, 60
Mercury, 140;
best times to find, 92
Meridian, Greenwich, 17;
longitude, 17;
observer's, 18
Milky Way, 66
Mira, 66
Mizar, 60
Moon, 162-189;
age, calculation for any date, 164;
anomaly, 166;
apogee, 165;
daily motion, approximate, 165;
eclipses of, 179, 186-188;
evection, 167;
horizontal parallax, 176;
latitude (ecliptic) calculation, 170;
longitude (ecliptic) calculation, 169;
nodes of orbit, 165;
"on its back", 91;
orbit of, 165;
perigee, 165;
phase of, 164;
position calculations, 165, 167-170;
position calculator, 170;
rotation on axis, 163;
terminator, 164;
time by, on sundial, 44

Navigation, 129;
using a celestial globe, 90
Nebulae, 62
Nocturnal, making and using, 53
Nodes of an orbit, 151

Occultations, 69, 178
Ophiuchus, constellation of, 72
Opposition of superior planets,140
Orbits, 136-140;
eccentricity, 151;
elements of, (table), 196;
measuring shape of, 158;
Moon's, 165;
nodes of, 151;
planetary, 139;
Pluto's, 152
Orion, constellation of, 63;
finding east and west by, 115;
Great Nebula in, 63

Parallax, horizontal, 176;
of Moon, 177;
of stars, 16
Pegasus, constellation of, 62
Penumbra, 181
Perihelion, 24;
longitude of, 151
Perseus, constellation of, 65
Photography, star trail, 21
Pisces, constellation of, 63;
Sun in, at equinox, 20
Piscis Austrinus, constellation of, 62
Planets, 136-161;
aphelion, 24;
conjunctions, 141;
distances of, 149, 196;

elongation of, 141;
inferior, 140;
opposition, 140;
orbital elements (table), 196;
perihelion, 24;
position calculators, 146;
position finding, 152-158;
retrograde apparent motion, 139;
superior, 138
Planetarium, papier maché, 78-81
Pleiades, 65
Plough, 57;
use with nocturnal, 55
Pluto, orbit of, 152
Polaris, 19, 57;
altitude of, 21;
position relative to true celestial pole, 60
Pole, celestial, 19;
terrestrial, 13
Pollux, 68
Position line, 133
Præ sepe, 68
Procyon, 66

Quadrant, making and using, 98-102

Regulus, 68
Retrograde motion of planets on celestial sphere, 139
Rigel, 63
Right ascension, 46-47;
relationship with sidereal time, 49

Sagitta, constellation, of, 74-76
Sagittarius, constellation of, 76
Scorpius, constellation of, 72
Semidiameter, 184
Shadows, measuring Sun's position by, 9;
length of planets', 179;
umbra and penumbra, 181
Sidereal time, 49;
approximate calculation of, 50-51;
astrolabe scales, 112;
globe scales, 87;
U.T., relationship with, 49-50;
by the nocturnal, 55;
right ascension, relationship with, 49;
by the stars, 51;
tables of, 194
Sirius, 66
Solar system, 140
Solstices, 14
Spica, 70
Stars, through the year, 57-77
"morning and evening", 140;
tables of names, magnitudes and positions, 200;
time by, 51;
trail photographs of, 21
Stereographic projection, 103;
in the astrolabe, 107;
scales of, 120
Sun, 24-28;
eclipses of, 179-186;
"mean", 26;
motion of around ecliptic, 26;
position of, calculations, 167;
position of, by shadow measurement, 9-13;
projection of image, 160;
time by, 13, 24-45
Sundials, 24-45;
analemmatic, 36;
angles between hour lines on, 33
Capuchin, 40;
equatorial, 28,
gnomon, 29;
hands, use as, 35;
horizontal, construction of, 29;
portable, combined analemmatic and horizontal, 40;

setting up in garden, 44;
style, 29
Superior conjunction, 141
Superior planets, 138
Symbols for astronomical terms, 192
Synodic month, 162
Synodic period, 154

Taurus, constellation of, 63
Telescope, projecting Sun's image with, 160
Terminator, 164
Time, British Summer, 10;
Greenwich Mean, 10;
local mean, 26, 49;
mean solar, 13;
sidereal, 49;
Universal, 26
Triangles, mathematical solutions of, 193
Triangulum, constellation of, 63
Tropics of Cancer and Capricorn, 14
Twilight, astronomical, civil and nautical, 109

Umbra, 181
Ursa Major, constellation of, 60
Ursa Minor, constellation of, 61

Variation, Moon's, 167
Vega, 73
Venus, 140,
finding in daylight, 102, 114;
position of, 142
Virgo, constellation of, 70

Year, days in, (table), 199;
leap years, 199

Zenith distance, 120
Zodiac, 63